Just thinking...

The Case for Intelligent Design

Donald L Johnson
Copyright 2018
Donald L Johnson
Helena Montana, USA
All rights reserved

α

I Wonder as I wander out under the sky ...

I remember those nights as a kid laying out in the backyard on a clear Montana night. Laying out in that mummy bag gazing in wonder at the night sky and ... just thinking ...

How many stars are out there? How big is the universe? Does it have an end? If it has an end ... then what's beyond? How did it get there? How did I get here? How can There be something that has no end? Are there answers?

I think most of us have had such nights, and for many of us such wonder continues ...

I'm one of those whose mind seems never to stop with the wondering ... I just can't shut it off – not that I want to.

So, this is a collection of thoughts, conversations and writings that I've collected over the years. Thoughts and study that go back many, many years. The pages to follow come from my thinking and study on these issues, from my blog and from internet dialogs I've had with various people, mainly commentators like myself who seem to be passionate about the same questions and issues. Many have quite the opposite views as me, but that's OK.

Ω

Dedication

This book is dedicated to those who dare go against the grain and against the intellectual, academic and professional tyranny of the Darwinian Evolution lobby. It is dedicated to those willing to follow evidence wherever it may lead, in the face of fierce, powerful and often times vicious opposition from those seeking to retain the power of their dominating world view.

This book is dedicated to those organizations and individuals seeking truth in science and philosophy:

- The Discovery Institute - https://evolutionnews.org/ In particular Casey Luskin, who befriended me and graciously reviewed a couple of my early articles.
- Uncommon Descent - https://uncommondescent.com/
- Institute for Creation Research - http://www.icr.org/
- To Pastor Richard Emery, who set me off on this journey so many years ago. Thank you, Dick.
- And others traveling along those oft perilous seas.

Thank you for looking at this book. If you have enjoyed it and wish to purchase copies, visit:

https://amazon.com/author/donjohnsonbooks

Or contact the author at:DonJohnsonDD682@live.com

Contents

Part I: The Case 1
  Just Thinking … 1
  Some Definitions 3
    Atheism: 3
    Creationism: 3
    Creation ex nihilo: 3
    Creation myth: 3
    Intelligent Design: 3
    Materialism: 3
    Atheism – My definition: 5
  Some contrasting thoughts 8
    Carl Sagan 8
    Neil deGrasse Tyson 10
    Richard Dawkins 12
    Stephen Hawking 15
    Bill Nye -- the science guy 17
    Richard Leontine PhD 19
  Shall we delve into some detail? 22
    Professional Evolutionists. Are they really all that smart? 24
    Looking Inside A Living Cell 29
    "The Unintelligent Designer" 32
    Evidence of design and the mechanisms involved 38
    Design -- Bottom Up or Top Down 48
    Mimicking a Neural Network? 55
    I've Grown Accustomed to Your Face 59
    A Short Conversation with a Real Scientist 63
    A word about "Deep Time" 74
    Machines everywhere! 77
    Jargon and Mumbo-Jumbo … where's the Evidence? .. 81
    Here is the Evidence 85
    Massively Complex Synchronicity – Part 1 86
    Massively Complex Synchronicity – Part 2 89
    Massively Complex Synchronicity – Part 3 100
    The Necessity for Simultaneity (i.e. The Creation Week) 104

Your Designed Body: truth and rejection	106
Part II: Evolution, Intelligent Design, Creation/Faith	107
Part III: Why Does All This Matter?	109
The Case of Richard Sternberg	112
The Case of Eric Hedin and Ball State University	117
A somewhat personal encounter with professional bigotry	138
Conclusion	140
About the author	141

Part I: The Case

Just Thinking ...

In this book I hold out the hope of showing that a world view centered around a designed and created universe is not only reasonable, but that such a view is the one that aligns with and makes sense with our everyday encounter with our universe and all that is within it – including you and me. Scientists of today as well as in years and centuries past have assigned the idea of "Rational Intelligibility" as an apt description of this view.

This view of a designed and created universe is a minority view these days, not held by many who are the shapers of the modern view of materialism which holds that everything we experience has come about through entirely natural means. In other words, Darwinian Evolution explains it all and we need seek no further for answers to the many mysterious and profound questions of life.

The popular purveyors of evolution such as Richard Dawkins, Neil deGrasse Tyson, Carl Sagan, Jerry Coyne, Steven Hawking, and others, dismiss our everyday encounters with design as – the *"appearance"* of design, but not real design, only an *"illusion"* of design – not the real experience of design experienced when you turn on your television to your favorite show, or talk to your friend on your iPhone.

If you are uncertain, or you hold to a world view of materialism and evolution as holding the ultimate answers, I would encourage you to step back for a moment and take a deep breath -- step out and take a risk to examine an alternative view. Examine the examples and thinking I give in the following pages, as well as seeking out your own examples. You can and should take this intellectual plunge initially in the privacy of your own mind -- take whatever time is appropriate for you to stir those brain cells into a stirring of questioning. You will indeed take a risk in doing this ... questioning your perhaps long-standing beliefs ... risking ridicule and censure, even from those close to you. It won't be easy.

Some Definitions

Atheism:
in a broad sense, the rejection of belief in the existence of deities. In a narrower sense, atheism is specifically the position that there are no deities. Most inclusively, atheism is the absence of belief that any deities exist. Atheism is contrasted with theism, which, in its most general form, is the belief that at least one deity exists.

Creationism:
the belief that the universe was created in specific divine acts and the social movement affiliated with it.

Creation ex nihilo:
the concept that matter comes "from nothing."

Creation myth:
a religious story of the origin of the world and how people first came to inhabit it.

Intelligent Design:
Refers to a scientific research program as well as a community of scientists, philosophers and other scholars who seek evidence of design in nature. The theory of intelligent design holds that certain features of the universe and of living things are best explained by an intelligent cause, not an undirected process such as natural selection.

Materialism:
A form of philosophical monism which holds that matter is the fundamental substance in nature, and that all phenomena, including

mental phenomena and consciousness, are identical with material interactions.

Atheism – My definition:

Atheism is much like locking yourself in a cave with no windows but has a door. Inside the cave there is a strong and long-standing taboo about that door: *Do not go near that door! Do not open that door, not even just a crack!* The type of science that an atheist uses searches in all of the corners of the cave looking for God (or gods if you prefer) and not finding him concludes that there is no such thing because he/she/it/them has not been found in the cave. There are also those who don't search at all -- but start off with the assertion and assumption that there is nothing beyond that door.

Then one day a brave soul decides to open the door to see what is outside. Outside of the cave he finds many wonderful, mysterious and interesting things, and then proceeds to seek and discover more about them and who or what might have created these things. He seeks answers more fully in ways spiritual (e.g. the Bible), in reflective thought and by science in examining the created things he has discovered outside of the cave.

But sadly, many still in the cave refuse to approach that door.

Since God (if there is such a thing) exists beyond time and space, He will never be discovered or proved using scientific methods.

Dare to open the door and see what lies beyond; you may be amazed at what you find.

The Night the Author Entered the Story [1]

Further to this cave analogy, I found this fascinating insight into C.S. Lewis's coming to faith in Jesus Christ in his book *Surprised by Joy*.[2]

I find C.S. Lewis's Hamlet illustration to be quite compelling. I shared it with high school students earlier today. In one sense, if Hamlet were to explore his world he wouldn't see any evidence of Shakespeare. He wouldn't find him in outer space, hiding behind a tree, or submerged in the depths of the ocean.

In another sense, however, Hamlet would see evidence of Shakespeare everywhere. He would be living in the world Shakespeare made. The existence of this world is entirely dependent upon its author.

But for Hamlet to know Shakespeare personally, intimately even, the author would have to write himself into the story. As Lewis said, "Hamlet could initiate nothing." For the two to meet, "it must be Shakespeare's doing."

This is the basic claim of the Christian narrative. There is an author to our story, to our world. If ever we would meet him, it will not be of

---

[1] The Night the Author Entered the Story (theolatte.com) https://www.theolatte.com/2017/12/the-author-entered-the-story/

[2] Surprised by Joy: The Shape of My Early Life: Lewis, C. S.: 9780062565433: Amazon.com: Books https://www.amazon.com/Surprised-Joy-Shape-Early-Life/dp/0062565435

our own initiation. He must act. He must write himself into human history. And he has. As the Apostle John says, "In the beginning was the Word . . . and the Word became flesh."

The scandalous claim of Scripture is that the Author entered the plot. He stepped over the threshold of time and space to be born in a seemingly insignificant town in the Middle East. The Creator of the world wrapped in a first-century diaper: phenomenal cosmic power in an itty, bitty living space. This is our best and only hope. This is joy. This is life. This is Christmas.

Joy to the world

The Lord has come

Let earth receive her King

Let every heart prepare Him room

And heaven and nature sing

Some contrasting thoughts

Carl Sagan

'The Cosmos is all that is or was or ever will be"

Then the Lord answered Job out of the whirlwind and said:

"Who is this that darkens counsel by words without knowledge? Dress for action like a man; I will question you, and you make it known to me.

"Where were you when I laid the foundation of the earth? Tell me, if you have understanding. Who determined its measurements— surely you know! Or who stretched the line upon it? On what were

its bases sunk, or who laid its cornerstone, when the morning stars sang together and all the sons of God shouted for joy?

"Or who shut in the sea with doors when it burst out from the womb, when I made clouds its garment and thick darkness its swaddling band, and prescribed limits for it and set bars and doors,11 and said, 'Thus far shall you come, and no farther, and here shall your proud waves be stayed'?   Job 38:1-11

A nice sound bite or bumper sticker.

But when I reflect on a statement such as Sagan's, and think of myself as saying that ... who would I be and where would I place myself?

I would have to be at a place where I can see the entirety of the universe ... all there is, even far beyond what our best telescopes can see. I would have to have solved the mysteries of the so-called Dark Matter and Dark Energy, consciousness, the origin, and development of life ... and more.

I would have to be witness to the Big Bang so as to see what preceded it, what happened during it, and how it unfolded at every turn.

I would have to have the attributes of Omniscience and Omnipresence.

But I am not at such a place, nor do I have such attributes.

What I would be ... is highly arrogant and self-centered. Perhaps highly educated, but also highly ignorant. I would put myself in the place of the very God I so wish to bring to naught. I would ascribe to myself the very attributes of that God.

I don't think I want to be that kind of person.

Neil deGrasse Tyson

"I think of, like, the human body, and I look at what's going on between our legs, There's like a sewage system and entertainment complex intermingling. No engineer of any intelligence would have designed it that way."

> For you created my inmost being you knit me together in my mother's womb. I praise you because I am fearfully and wonderfully made; your works are wonderful, I know that full well. My frame was not hidden from you when I was made in the secret place, when I was woven together in the depths of the earth. Your eyes saw my unformed body; all the days ordained for me were written in your book before one of them came to be. *Psalm 139:13-16*

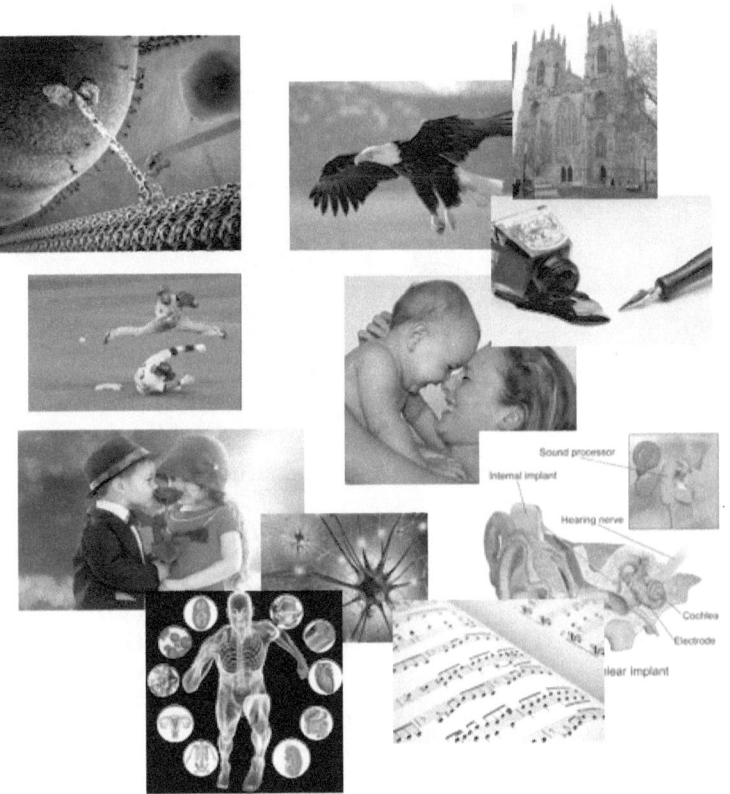

No engineer of any intelligence *__could have__* envisioned, designed, engineered and manufactured such an amazing machine as the human body. The vast variety of purpose, goals and functionality demonstrated by the human mind and body is simply breathtaking.

Richard Dawkins

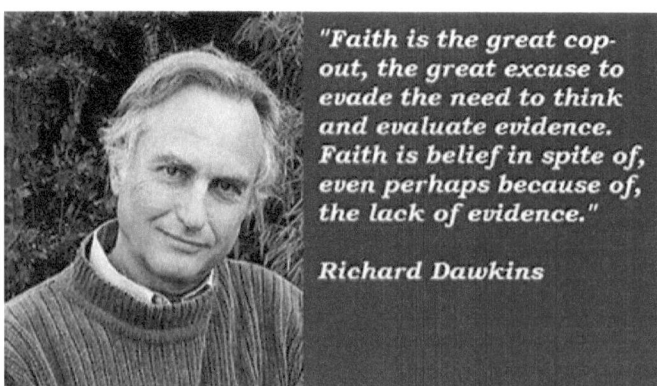

"Biology is the study of complicated things that give the appearance of having been designed for a purpose."

For My thoughts are not your thoughts, nor are your ways My ways, says the Lord. For as the heavens are higher than the earth, so are My ways higher than your ways and My thoughts than your thoughts. For as the rain comes down, and the snow from heaven, and do not return there but water the earth and make it bring forth and bud that it may give seed to the sower and bread to the eater, so shall My word be that goes forth from My mouth; it shall not return to Me void, but it shall accomplish that which I please, and it shall prosper in the thing for which I sent it. **Isaiah 55:8-11**

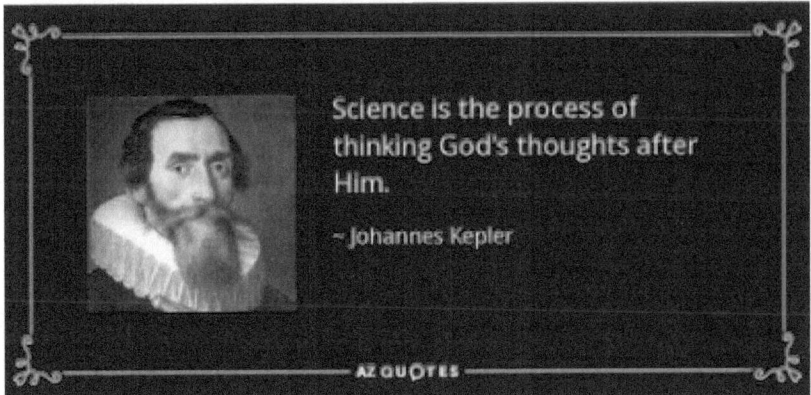

Kepler was not alone in believing that nature was a book in which the divine plan was written. He differed, however, in the original manner and personal intensity with which he believed his ideas to be embodied in nature. One of the ideas to which he was most strongly attached—the image of the Christian Trinity as symbolized by a geometric sphere and, hence, the visible, created world—was literally a reflection of this divine mystery (God the Father: center; Christ the Son: circumference; Holy Spirit: intervening space). One of Kepler's favorite biblical passages came from John 1:14: "And the Word became flesh and lived among us." For him, this signified that the divine archetypes were literally made visible as geometric forms (straight and curved) that configured the spatial arrangement of tangible, corporeal entities. Moreover, Kepler's God was a dynamic, creative being whose presence in the world was symbolized by the Sun's body as the source of a dynamic force that continually moved the planets. The natural world was like a mirror that precisely reflected and embodied these divine ideas. Inspired by Platonic notions of innate ideas in the soul, Kepler believed that the human mind was ideally created to understand the world's structure.

Johannes Kepler, a man of faith, was a key player in a profound change in the tide of human thought: *the scientific revolution.*

Perhaps a greater contributor to science and knowledge than Richard Dawkins.

Stephen Hawking

" ... tiny quantum fluctuations in the very early universe became the seeds from which galaxies, stars, and ultimately human life emerged ... ."

He alone stretches out the heavens and treads on the waves of the sea. He is the Maker of the Bear and Orion, the Pleiades and the constellations of the south. He performs wonders that cannot be fathomed, miracles that cannot be counted.   Job 9: 8-10

But who has seen and examined the very early universe -- and what is the nature of the tiny quantum fluctuations?

Might not the following be more reasonable?

'In the beginning God created the heavens and the earth ... '

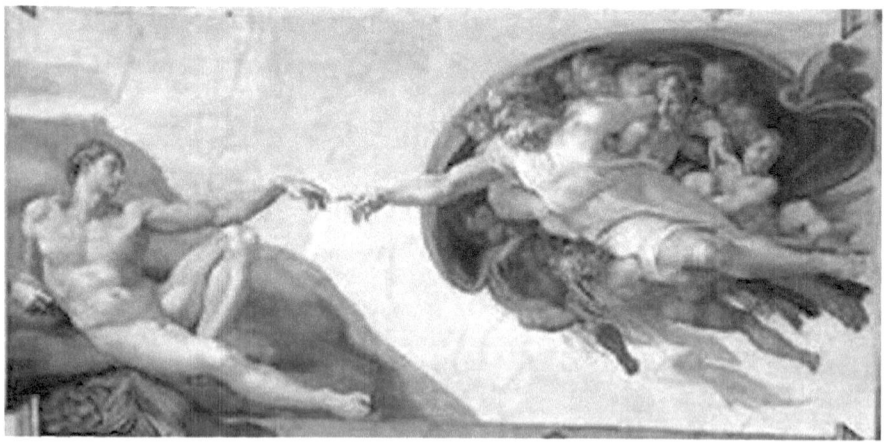

Bill Nye -- the science guy

I'm insignificant. ... I am just another speck of sand. And the earth really in the cosmic scheme of things is another speck. And the sun an unremarkable star. ... And the galaxy is a speck. I'm a speck on a speck orbiting a speck among other specks among still other specks in the middle of specklessness. I suck.

"In the past God spoke to our ancestors through the prophets at many times and in various ways, but in these last days he has spoken to us by his Son, whom he appointed heir of all things, and through whom also he made the universe. The Son is the radiance of God's glory and the exact representation of his being, sustaining all things by his powerful word. After he had provided purification for sins, he sat down at the right hand of the Majesty in heaven." Hebrews 1:1-3

What the Science Guy says is absolutely true ... but only when the material is considered. We are indeed insignificant; as a solar system, as a planet and as individuals -- and the size of these specks will only diminish as we explore and discover more and more of the universe.

However, for those of us willing to include God in our thinking and into our souls, an entirely different vista opens before us ... . Again, listen to these words written thousands of years ago about the "insignificance" of the earth:

"For this is what the Lord says— he who created the heavens, he is God; he who fashioned and made the earth, he founded it; he did not create it to be empty, but formed it to be inhabited— " Isiah (45:18)

And as to you and me being an insignificant speck, again read these words:

> "Even the darkness is not dark to You, And the night is as bright as the day. Darkness and light are alike to You. For You formed my inward parts; You wove me in my mother's womb. I will give thanks to You, for I am fearfully and wonderfully made; Wonderful are Your works, And my soul knows it very well.... " **Psalm 139:14**

Richard Leontine PhD

'Our willingness to accept scientific claims that are against common sense is the key to an understanding of the real struggle between science and the supernatural. We take the side of science in spite of the patent absurdity of some of its constructs, in spite of its failure to fulfill many of its extravagant promises of health and life, in spite of the tolerance of the scientific community for unsubstantiated just-so stories, because we have a prior commitment, a commitment to materialism.

It is not that the methods and institutions of science somehow compel us to accept a material explanation of the phenomenal world, but, on the contrary, that we are forced by our a priori adherence to material causes to create an apparatus of investigation and a set of concepts that produce material explanations, no matter how counter-intuitive, no matter how mystifying to the uninitiated. Moreover, that materialism is absolute, for we cannot allow a Divine Foot in the door."

"For the wrath of God is revealed from heaven against all ungodliness and unrighteousness of men, who suppress the truth in unrighteousness, because what may be known of God is manifest in them, for God has shown it to them. For since the creation of the world His invisible attributes are clearly seen, being understood by the things that are made, even His eternal power and Godhead, so that they are without excuse, because, although they knew God, they did not glorify Him as God, nor were thankful, but became futile in their thoughts, and their foolish hearts were darkened. Professing to be wise, they became fools ... "

Might not *Richard Leontine, the* PhD, be that one sitting on the rock lecturing the other cave dwellers on the "fact" that the bright light outside the cave in only an illusion ... there is no evidence that such a "Sun" exists ... don't go near the door. There's no "out there" out there!!!

Still thinking ...

Shall we delve into some detail?

Among the staples of the materialistic world view, and of Darwinian evolution in particular, are the notions that everything in nature derives from totally material means ... no supernatural events or personages are allowed ... or even considered.

Further, in the naturalist world view, evolution is non-goal directed and has no purpose. Life becomes more complex simply through a long series of random mutations from which "natural selection" keeps those mutations beneficial to the survival of the life form at any given point in time, and presumably adds them to the ever-evolving mix, thus assuring the "survival of the fittest." This is called "***deep time***" and is the recipe for the vast diversity of life we see on this home we call Earth.

However, there seems to be a small problem with this scenario. Wherever we look at life, be it at the cellular level or up through the various internal organs of a body, and indeed the completed body itself, what we see is purpose. Let me repeat ... we see purpose ... and goals.

And we see **"*Massively Complex Synchronicity.*"**

I hope some of the examples and articles included in this book will allow you to see purpose and goals fulfilled. I also hope these examples will cause you to think and ask questions such as *"how could these purposes arrive on the scene through the Darwinian process?"*

So, let's begin this journey.

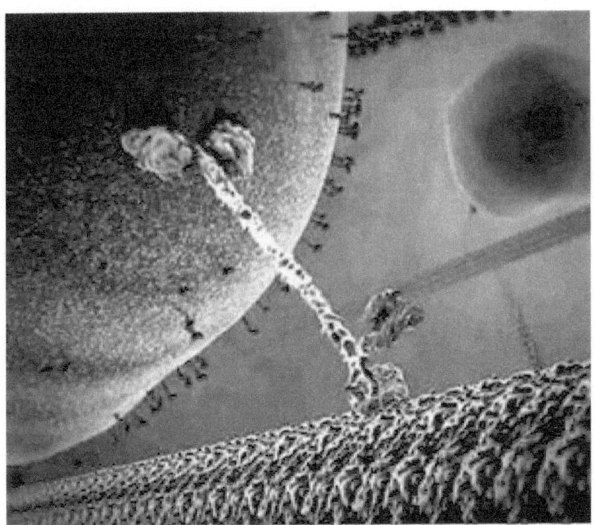

The Kinesin Motor depicted above is a truck, moving cargo from one place in a cell to another along a tubular roadway -- and the roadway is constructed just in front of the truck as needed, and is disassembled as the truck and its load moves along.

The beauty of a well-executed double play. Purpose and goals on display before thousands of cheering fans.

Professional Evolutionists. Are they really all that smart?

Prologue

A couple of events have come to my attention after I wrote this essay, and I'm wondering if there could be any connection.

First is a recent Texas textbook controversy to include only evolution. It seems there is invariably a flurry of anti-creation activity in the popular press coincident with such decisions.

Second is the mocking of Governor Rick Perry's comments on creation and evolution in Texas and calling into question Michelle Bachman's faith as related to her governance qualifications.

Just wondering!

I don't think they are all that smart, and they would like to dumb-down you as well. Or they would like to deceive you and make you as stupid as they are.

The pattern is consistent and very old; I have been seeing it in the popular literature for over 30 years now, and it seldom varies from the script. A bold statement up front claiming the settled scientific truth and fact of evolution then followed by a long series of speculations and suppositions supported by little if any scientific fact.

A couple of illustrations to make my point:

You may have seen a Discovery Magazine publication *"Evolution"* with the subtitle *"From Ooze to Us"* Let's browse through the Table of Contents to get a sense of what is contained within (**emphasis mine**):

- The Slime Years. 4 Billion to 2 Billion Years Ago.

- A Cold Start. Freezing temperatures **might have** jump started the chemistry of life.
- Did life start with a Virus? .... **could have been** the very foundation of Biology.
- The Living Fossils, some of today's organisms **appear almost untouched** ....
- Before the womb. A controversial theory .... **asserts** that earliest mammals **probably crawled out**

Do you see what is happening here? Solid scientific observations and evidence is replaced by *"might have ..."*, *"could have been ..."*, *"probably ..."* and other similar speculations. Sloppy and deceptive is what it is!

If you actually read the articles, and many others of a similar ilk, you will get more of the same, and little if any solid science.

Then there is an issue of Scientific American (July 2011) and an article on the *"Evolution of the eye"* by Trevor d. Lamb. Again, the article makes a bold claim of scientific fact, and claims to demolish Creationist and Intelligent Design arguments about the eye. The article then follows the old familiar pattern. I've highlighted many of the catch phrases (*my emphasis added*) in the article which illustrate the shallowness of the supposed research:

- ... these findings **put the nail in the coffin** of irreducible complexity and **beautifully support** Darwin's idea. ...
- **Around a billion years ago** simple multicellular animals diverged into two groups: ...
- The bilateria themselves **then diverged** ...
- This burst of evolution **laid the groundwork** for the emergence of our complex eye.
- ... during the Cambrian explosion two fundamentally different styles of eye arose. The first **seems to have been** a compound eye ...
- ... such visual ability **may have given** ...

- Hence, *as body size increased, so too, did* the selective pressures favoring ...
- This pattern **suggests** ...
- ... which ***probably resembled*** modern day ...
- We reasoned that ... ***must therefore*** have still deeper roots
- ... although it **has apparently** not evolved ...
- These **striking similarities** ... are far too numerous ...
- ... ***must have*** been present ...
- Unfortunately, there are no living representatives ...
- But we found clues ...
- Observations of hagfish behavior **suggest** ...
- ... this ancestor ***presumably had*** ...
- This persistence ***suggests that*** ...
- ... ***could thus*** throw light on ...
- Hints about the role of the hagfish eye ... ***Perhaps then*** ...
- ... first served ... ***and only later*** evolved ... Studies ... support this notion.
- We can with caution ...
- ... to inform our reconstruction of how the eye evolved.
- ... exhibits ***telltale clues*** ...
- ... one ***would expect to see if*** the vertebrate retina evolved
- It therefore seems ***entirely plausible*** ...
- ... ***represents a holdover*** from a period in evolution ...
- ... which evolved independently ...
- ... early in the evolution of ... ***a change occurred*** ...
- ... **may have been** able to ...
- We postulate that ...
- ... or – had evolved in an ancestor of ...
- It seems likely that ...
- ... ***could have*** easily evolved ...
- ... ***may have*** risen ...

- ... in a geological instant.
- ... because they **presumably had** ...
- Thus, there **would have been** ...
- ... must have been present ...

The author then concludes this hammering the final nail into the coffin of intelligent design/creation as follows: "*The design of our eye is not intelligent – but it makes perfect sense when viewed in the bright light of evolution.*"

Hogwash!!! It does no such thing. All it shows is the ignorance and bigotry of the author, a supposed scientist. I urge you to seek the truth in this matter yourself, don't trust me, and don't trust supposed experts such as Mr. Lamb. Breaking away from this evolutionary propaganda may be difficult for you as it was for me a number of years ago. Evolutionary thinking of this nature is like an addiction, and addictions are hard to break free from; ask a recovering alcoholic, or one trying to quit smoking. But it can be done. There is much good scientific research and data and evidence available out there that supports the Creation/Intelligent Design model of life. And start using your own common sense and personal observations as to the beautiful design attributes of what you see, hear, feel, and smell around you.

You would think that major publications targeting the public at large would be able to come up with convincing evidence and facts on evolution. I consistently find them greatly lacking in this area, and this from a magazine that touts on its cover "*winner of the 2011 National Magazine Award for General excellence*".

What's interesting about the SA article is that this same magazine published an incredibly fascinating article "The Movies in Our Eyes" (April 2007). This article shows legitimate research into the incredible capability of the eye and brain. The authors liken the capabilities of the eye to a system having 12 cameras, each having a visual and spatial specialization (with some overlap between the various cameras) giving us the wonderful gift of vision. The article sticks to observed fact and phenomena and avoids entirely the nonsensical speculations of the articles referenced above. There are no speculations as to how this marvelous visual system came into being, just a straightforward presentation of a scientific investigation.

After reading this article, I was impressed in how strong a case was made for the eye being a designed and created system, and with absolutely no mention of creation or intelligent design (nor of evolution).

It seems as if somewhere along the road to PhD, some of these folks take on a strong dose of stupidity rather than the strong dose of wisdom you would expect from an advanced degree from a prestigious university. But then again, I'm just the son of an immigrant TV repairman from Butte Montana, what could I possibly know?

"Deep Time" evolution

Looking Inside A Living Cell

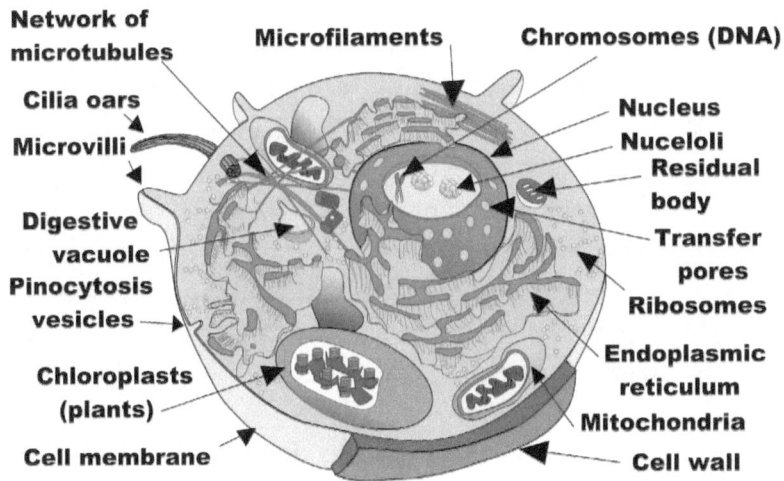

"To grasp the reality of life as it has been revealed by molecular biology, we must magnify a cell a thousand million times until it is twenty kilometers in diameter and resembles a giant airship large enough to cover a great city like London or New York. What we would then see would be an object of unparalleled complexity and adaptive design. On the surface of the cell we would see millions of openings, like the portholes of a vast space ship, opening and closing to allow a continual stream of materials to flow in and out. If we were to enter one of these openings we would find ourselves in a world of supreme technology and bewildering complexity. We would see endless highly organized corridors and conduits branching in every direction away from the perimeter of the cell, some leading to the central memory bank in the nucleus and others to assembly plants and processing units. The nucleus of itself would be a vast spherical chamber more than a kilometer in diameter, resembling a geodesic dome inside of which we would see, all neatly stacked together in ordered arrays, the miles of coiled chains of the DNA molecules. A huge range of products and raw materials would shuttle along all the manifold conduits in a highly ordered fashion to and from all the various assembly plants in the outer regions of the cell.

We would wonder at the level of control implicit in the movement of so many objects down so many seemingly endless conduits, all in perfect unison. We would see all around us, in every direction we looked, all sorts of robot-like machines. We would notice that the simplest of the functional components of the cell, the protein molecules, were astonishingly, complex pieces of molecular machinery, each one consisting of about three thousand atoms arranged in highly organized 3-D spatial conformation. We would wonder even more as we watched the strangely purposeful activities of these weird molecular machines, particularly when we realized that, despite all our accumulated knowledge of physics and chemistry, the task of designing one such molecular machine – that is one single functional protein molecule – would be completely beyond our capacity at present and will probably not be achieved until at least the beginning of the next century. Yet the life of the cell depends on the integrated activities of thousands, certainly tens, and probably hundreds of thousands of different protein molecules.

We would see that nearly every feature of our own advanced machines had its analogue in the cell: artificial languages and their decoding systems, memory banks for information storage and retrieval, elegant control systems regulating the automated assembly of parts and components, error fail-safe and proof-reading devices utilized for quality control, assembly processes involving the principle of prefabrication and modular construction. In fact, so deep would be the feeling of deja-vu, so persuasive the analogy, that much of the terminology we would use to describe this fascinating molecular reality would be borrowed from the world of late twentieth-century technology.

What we would be witnessing would be an object resembling an immense automated factory, a factory larger than a city and carrying out almost as many unique functions as all the manufacturing activities of man on earth. However, it would be a factory which would have one capacity not equaled in any of our own most advanced machines, for it would be capable of replicating its entire structure within a matter of a few hours. To witness such an act at a magnification of one thousand million times would be an awe-inspiring spectacle."

Credit: Michael Denton PhD., Evolution: A Theory in Crisis, pg.328

"The Unintelligent Designer"

Several books have come to my attention that can be placed in the category of *"If I were God, I wouldn't have done it that way! Therefore, there is no god"* There are others as well such as Rosa Rubicondior.

Here are a few snippets from the book *"The Not-So-Intelligent Designer"* by Abby Hafer:

"Why do men's testicles hang outside the body? Why does our appendix sometimes explode and kill us? And who does the Designer like better, anyway- us or squid?"

Dr. Abby Hafer (doctorate in zoology from Oxford University and teaches human anatomy and physiology) argues that the human body has many faulty design features that would never have been the choice of an intelligent creator. She also points out that there are other animals that got better body parts, which makes the Designer look a bit strange.

ID critics such as Dr. Hafer and Glenn Branch of the National Center for Science Education (NCSE) claim to expose the fallacy of Intelligent Design by showing that, when examined in detail, biological systems are anything but intelligently designed. They show no signs of a plan and are quite ludicrously complex for whatever can be described as a purpose.

They claim that the Intelligent Design movement relies on almost total ignorance of biological science and seemingly limitless credulity in its target marks. Its only real appeal appears to be to those who find science too difficult or too much trouble to learn yet want their opinions to be regarded as at least as important as those of scientists and experts in their fields.

I must beg to differ on Dr. Hafer's and Rosa Rubicondior's view of this so called *"Not-so-Intelligent Designer"* as well as the views of Glenn Branch of NCSE. (Search Amazon to learn more about these books.)

What they see as "quirks and kinks, the makeshift solutions and haywire failures of human biology," many see as an elegant and quite magnificent design with an amazing and far-ranging menu of capabilities. Let me suggest an exercise that the authors, Mr. Branch and others can easily accomplish, and I believe you may see my point.

Take an evening out and partake in one of those wonderful choral and orchestra performances taking place all around the world at any given time – I would recommend Handel's Messiah for this exercise.

As you are watching and listing to this amazing musical performance, I would like you to notice and watch a number of things very carefully.

First the hands – the hands and body motions of the conductor, as well as his facial expressions and body movements as he leads the choir and orchestra through this magnificent musical piece.

Continuing with the hands — watch the hands, and in particular the fingers of the orchestra members as they travel across the various instruments – the string section, the brass section, the woodwinds — the piano. Watch carefully as their hands precisely match the direction given by the conductor. Watch as the fingers subtly, and at times strongly tease the music from their instruments.

And note the various musical instruments — envisioned, designed and created by many beautifully designed and created hands and fingers.

Next the choral voices – listen as these beautiful voices blend together perfectly with the orchestra and watch the faces and mouths as they blend perfectly with the hands of the conductor and with the orchestra.

Next listen and pay attention to your own reaction as the message of the words and music bring excitement and inspiration into your heart and soul.

As you leave the concert hall, take time to look at the building and its architecture and artistry. Again, the hands, arms, legs, and mind of those artisans designed those arches, paintings and sculptures you admire so much.

And when you get back home in bed, ponder over the creation of the musical score of the "Messiah." Imagine Handel hovering over his desk and the paper taking on lines and musical symbols and words. Imagine him going back and forth over that manuscript as he goes to and from the scriptures that are inspiring him. Imagine the music that is building inside his head as he creates this masterpiece.

No – the human body is not the "quirks and kinks, the makeshift solutions and haywire failures, of human biology," but is something far more splendid and wonderful.

Next, I would suggest a couple of sporting events for Dr. Hafer and Glenn Branch.

First to a major league baseball game where they can witness the flawless execution of a double play. Beginning with the pitcher placing the ball across home plate at 90+ mph. We then see the batter follow that fast moving and curving baseball with his eyes, calculating where it will be as it passes into the strike zone where he can then attempt to hit it with his hand/eye coordinated swing. Then we see the shortstop field the fast-moving ball after anticipating and calculating where it will enter his glove. He then shovels it off to the second baseman who tags the runner out while leaping over the runner, and then a quick and precise throw to the first baseman who steps on first base for the second out of the double play.

Next, we go to an NBA basketball game where we witness the continual back and forth of finely tuned, trained and coordinated athletes showcasing example after example of what these well-designed machines are capable of.

Next, we are off to an NFL football game where we witness precision in the well-designed human body of a quarterback throwing the football with precise accuracy to a fast moving and maneuvering receiver who stretches his body out to execute a fingertip catch as he passes the goal line in front of a defender for a touchdown.

No – the human body is not the "quirks and kinks, the makeshift solutions and haywire failures, of human biology," but is something far more splendid and wonderful.

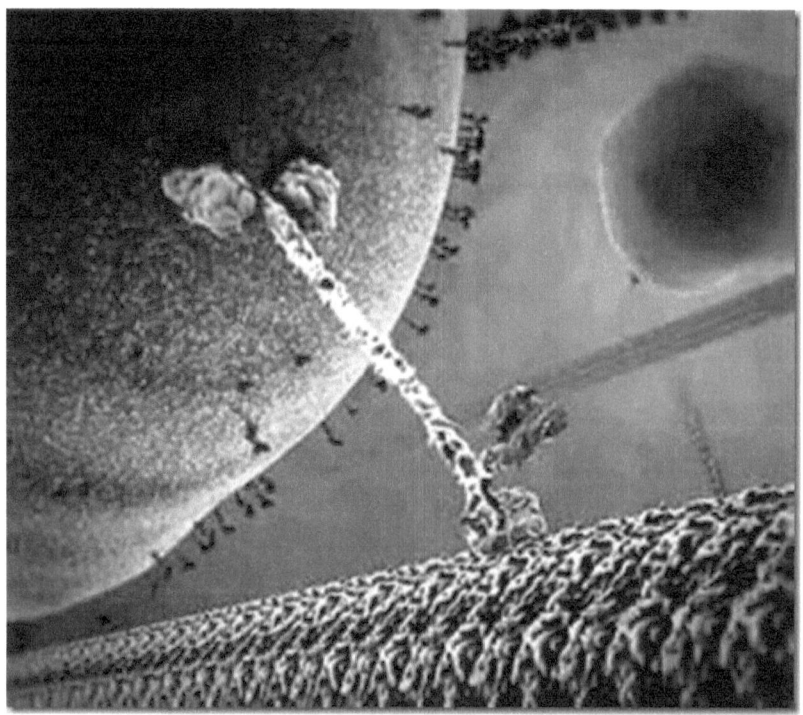

Next, we travel into the applied biological science of medicine and medical research and invite teachers and Mr. Branch to read and study the extensive articles that a Dr. Howard Glicksman has compiled on the intricacies and design of many aspects of the human body. We see this compilation of science reporting at the Discovery Institute web site. This series contains at least a half dozen articles on blood pressure alone, and how it is controlled within the human body. Note that this science reporting by Dr. Glicksman is seen at the Discovery Institute web site http://www.evolutionnews.org - an ID site - and not on the pages of the National Center for Science Education. I have been following the NCSE site as well as the Discovery site and others for years now, and what I find is that good science reporting like I describe above is found often and on a regular basis there, whereas seldom - approaching never - is an any science reported by NCSE. I find that very interesting and troubling, and thus would offer a caution to teachers to view NCSE with a great deal of skepticism, and especially these books which are little more than a hit piece on those of us who differ with the atheistic stance and mission of NCSE.

No - the human body is not the "quirks and kinks, the makeshift solutions and haywire failures, of human biology," but is something far more splendid and wonderful.

The activities I have described are direct evidence of goal directed purpose and design - evidence of an Intelligent Designer.

Hmmm ... let me get back to you on that?

Evidence of design and the mechanisms involved

The following is outfall from a comment I made to an article at the National Center for Science Education (NCSE). My comment prompted quite a response, in particular from Ian Nicholas.

Here is my initial comment:

"It is claimed on this site that Intelligent Design is not science, and that evolution is undisputed fact. Then I would like to see a response to the fact that life is full of purposeful machines such as the Kinesin motor as described here and at the following link:

... Kinesin offers a fascinating example of undiscovered information in action. What programs and machinery are required to assemble the structure and function of Kinesin? What information is needed for Kinesin to achieve its "runtime" functions? How does Kinesin know where to go to pick up a load, what load to pick up, what path to take, and where to drop its load? How does it know what to do next? All this functionality takes information, which must be encoded somewhere.

Indeed, the level of complexity is monotonically increasing, with no end in sight. ... "as highlighted in this article "Evolution's Grand Challenge" by Steve Laufmann at: http://www.evolutionnews.org/.../evolutions_gran097591.html and here to see an animation of these amazing machines in action: http://www.evolutionnews.org/.../did_kinesins_aro85951.html

Fossils may be interesting and fascinating, but they are dead. NCSE stays well clear of the actual science that is going on in studying, understanding and reporting on the "live" designs and machines that are all around and within us. Take a look, you might be surprised and awed by it all. "

And here are several comments among many, from Ian, challenging the idea of "purposeful machines" and "evidence."

"Your second link is also creationist apologetics which makes rather grandiose claims of "purposeful machines" in spite of the fact that you can't tell us the "purpose" of any of these things. "

"Evidence is irrelevant to your position."

"And if you have evidence of design then surely you should be able to tell us exactly what that evidence is and what the mechanisms are that are responsible?"

What follows is my response to some of the challenges that Ian has thrown out. I have tried to post it directly to the NCSE Facebook page, but for some reason my comments failed to upload.

"My goodness you were sure busy yesterday. I feel I must owe you an apology for causing you to spend so much time and energy responding to me. But then I realize that what I am seeing is your passion, and that is a good thing. I've been a passionate person for many years about quite a range of things, so I relate quite well to the passions of others.

You mentioned the so called "Gish gallop" earlier, and I guess that's what I'm seeing in your flurry of responses. I have followed the Institute for Creation Research for years, so I am familiar with Dr. Dwayne Gish. I know he was noted as quite a formidable debater although oddly enough I never have heard even one of his debates, but I can infer what is meant by the term "Gish gallop." He attended the same church in San Diego as we did, and I would see him and his wife there on a regular basis and actually met him personally.

I see you have read some of my blog posts and I thank you for that. These posts are from an interested lay person and not from a PhD practitioner in the fields of life science or medicine – and I gather you are in the same category as an interested lay person. You can see from my writings that I touch on many aspects of life and how it perhaps developed and flourished here on planet Earth. So perhaps I have addressed a fair number of your topics and criticisms of my world view, and I will not take the time to respond to each of your topics in turn. Many of my essays have been directed towards pointing the reader to the work and research of others who have a great deal of expertise and hands on experience with biological systems such as the human body.

But there are a few of your statements and criticisms that I will address here, and for the most part I will restrict my comments to the human body.

First is evidence. I have covered that previously, and on many occasions on my blog, and have pointed out that evidence of design is all around and within us. That evidence being the existence of purposeful machines and systems found at all levels of a human body from the cell containing its many machines, to the many purposeful organs each of us have, as well as the completed body itself.

One such machine (of many) I find particularly interesting is the Kinesin motor. As you can see from the animations of Kinesin, its main "purpose" is to transport cargo from one place in the cell to another. A fascinating adjunct to this machine is the roadway that the Kinesin traverses along its journey ... this roadway as it turns out is constructed in a "just in time" fashion ahead of the Kinesin and its cargo and is deconstructed after it is used.

I've been seeing these animations over the years, but always had the questions of how accurate they are in depicting what is actually going on in the cell and is the instrumentation these days actually good enough to see what is happening. Another question I have had is why the animations in the first place and not just the actual video taken of the biology itself.

Well, over the past year or so I've had a couple of opportunities to ask those questions of a couple of scientists actually working in the field. One was a researcher at the Perelman School of Medicine at the University of Pennsylvania, and the other was an MIT researcher at a lecture I attended at Yale this past year. The answers to my questions were the same from both gentlemen – yes the instrumentation is that good that it can see the kinds of things that the animations depict. And the animations are quite accurate in replicating the activities of such things as the Kinesin motor. As to the reason for the animations — as I recall, the reason is twofold; one is to isolate the object under investigation from all of the other busyness going on in the cell, so it can be studied somewhat in isolation, secondly, the animations being computer programs avail the researchers the opportunity to manipulate and tweak the object in their investigations.

Kinesin is but one of a number of machines that have been identified in the cell as having purpose – and yes, the researchers do refer to them as machines. Kinesin can be likened to the freeways and roadways we see and use on a daily basis – functional and designed with the purpose of transporting cargo.

The photos nearby show the raw data from instruments used to exam the internal workings of a cell. I happened to see these actual photographs on prominent display on walls at the Philadelphia airport, and they bore the name of the researcher at the Perelman School of Medicine at the University of Pennsylvania I mention elsewhere.

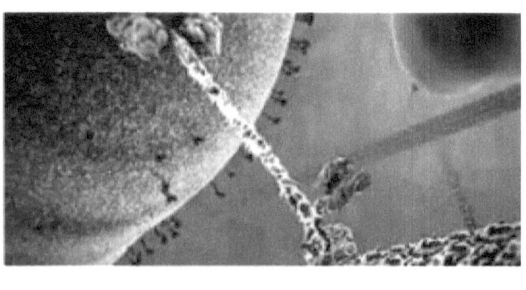

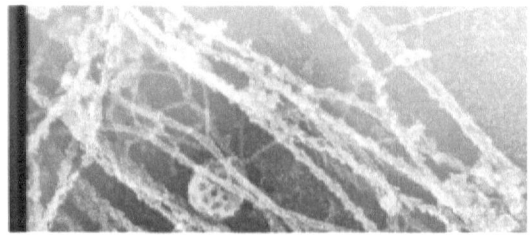

Another mechanism within the cell is that which replicates DNA strands. I don't have an animation at the ready for you, but I'm sure a search will find one quickly and I encourage you to find one and watch. An interesting and purposeful part of this DNA replication is what is referred to as Quality Control (QC) where a machine-like mechanism traverses the newly replicate DNA strand looking for errors in the transcription. When an error is found, the replication is paused while an attempt is made to repair the problem or failing that the new strand is killed off so as not to cause problems downstream.

So, there is much identified "purpose" within each of the trillions of cells making up our body. I hope this addresses your question "What's the "purpose" of kinesin then? And how'd you figure that out?" I didn't figure it out Ian, those much smarter than I figured it out.

Then we can move up from the cellular level to the various organs we all have. The major organs and systems within the human body each have specific purpose(s), and I won't dwell on each of these organs and functions.

But I would like to pass on to you a purposeful part of the circulatory system that I leaned about just yesterday. Here is a snippet from the article:

"... There are sensors located in the main arteries directly supplying blood to the brain, which can detect wall distension. These are the baroreceptors, which by sensing the stretching within the arterial walls are able to detect the arterial blood pressure. They are a type of mechanoreceptor that senses movement, in contrast to the chemoreceptors which detect chemicals like oxygen, carbon dioxide and hydrogen ion. The baroreceptors send their data on the blood pressure by way of nerves to the brain. The brain integrates this information, and if the blood pressure is too low, it causes the release of more norepinephrine and epinephrine from the sympathetic nerves. By attaching to specific receptors, increased sympathetic stimulation affects all three of the factors mentioned above, which makes the blood pressure rise. ... "

So here we have an answer to another of your statements/questions:

'... if you have evidence of design then surely you should be able to tell us exactly what that evidence is and what the mechanisms are that are responsible? '

This evidence being the very well-regulated circulation of blood – and the mechanisms are in part the baroreceptors and mechanoreceptors identified in the referenced article.

So, I find it interesting Ian that somehow the so-called undirected mechanism of Natural Selection somehow results in very directed and specifically functional and purposeful end products – from the cellular level on up to the completed body itself which is functionally capable of achieving a wide variety of functional tasks from writing the great American novel — to holding a baby- to executing that perfect double play in baseball.

Yes Ian ... we do indeed see purpose in the things that make up life – at all levels.

And where do we find such marvels reported? Not in this NCSE site where topics such as "Computational Biology", "biomimetics" and "Systems Biology" are seldom (I would venture to say never) mentioned let alone seriously reported on. No Ian, what you find here is an abundance of stories of fossils, the Grand Canyon and court cases but very little of what is happening in the laboratories of places such as the Perelman School of Medicine at the University of Pennsylvania.

As a layman where do I find an abundance of good reporting in cellular level biology and other fascinating stories of cutting edge research?, I find many of these reports at places like

http://www.uncommondescent.com/

and
http://www.evolutionnews.org/2015/09/controlling_blo099611.html

You might want to spend a little more time in such places Ian ... they could be dangerous to your world view, but very rewarding to your curiosity in the long run.

I do thank you for your responses, Ian. I don't get angry or defensive at what you write, but rather take them as opportunities for communication and sharing and contrasting often contrary and conflicting world views."

Still Thinking ...

Gosh ... I don't know. But my professor said ...

Design -- Bottom Up or Top Down

One of the fundamental fault lines separating Intelligent Design from Darwinian evolution is in how design (or the appearance of design) is expressed.

All human designed products are recognized to have been designed and built top-down -- first the vision (the goal), then designs and plans of the various component parts -- followed by the construction of the product. This process is readily seen daily wherever we look.

Evolution, on the other hand, since it is claimed to be non-goal directed and without purpose, must construct its products bottom-up through the endless tweaking of mutations and natural selection. No one has ever seen this process in action, but the process is inferred by the end result -- the product.

Which is it -- Bottom Up or Top Down?

Appealing to common sense and intuition, take a look at a few very sophisticated and complex human designed and manufactured systems.

What do you see in these two examples in the following pages?

We see some sort of chemical or petroleum processing factory. But in looking at the first picture a bit closer, we see two separate but mutually cooperative items -- both complex and both obviously designed and engineered for specific purpose and functionality. The one that first caught my eye was the massive array of interconnecting pipes and various machines that look like they could be manufacturing devices of some sort related to and interconnected to the pipes. At the bottom is what looks like a structure holding all this stuff in place lest it drift away into oblivion.

Obviously, an example of an intelligently designed unit representing much information related to its design, engineering and manufacture. Though it may not be obvious to readers of this book, it also represents an image of its purpose and functionality.

Then we see a massive ship carrying the mass of pipes machine. Again, the ship is specifically designed to carry its cargo to a particular place where it will be off loaded.

The second picture likewise shows a very complex human designed system, most likely a petro-chemical plant of some sort. Its identification is not important in this context -- the point to be made is that we can immediately recognize design when we see it.

These are the sort of "top down" designs we see all around us in the things we humans build.

Which is it -- Top Down or Bottom Up?

Continuing with an appeal to common sense and intuition, take a look at these very sophisticated and complex schematic diagrams in the following pages produced by researchers at a pharmaceutical company.

What do you see?

These schematics are the results of what's known as reverse engineering. This is a common engineering process where designs are teased out of often complex systems where there is little or no documentation describing the functionality of the system being examined – no blueprints, no specifications, no operator's manual and often no-one to talk too. It is often necessary to reverse engineer something in order to reproduce it, make modifications and improvements to it, and to troubleshoot when things go awry in something not well understood.

As a lifelong professional software developer, reverse engineering was a constant and ongoing effort to understand a complex piece of software written by others, sometimes long ago.

Other than an ideological need to deny obvious designs when an "apparent" design in nature is encountered, the prudent common sense and scientific path is to treat such "illusions" as actual designs and then seek to understand them in terms of design. This involves searching out functionality, purpose, and goals within the entity under study. I see this process loud and clear in the schematics and diagrams shown here.

And by the way, this process -- of seeking functionality and purpose -- this reverse engineering, runs almost completely counter to modern day evolutionary biology – but it succeeds.

Its name is Intelligent Design and is a top-down process.

I think I might see what's wrong here ...

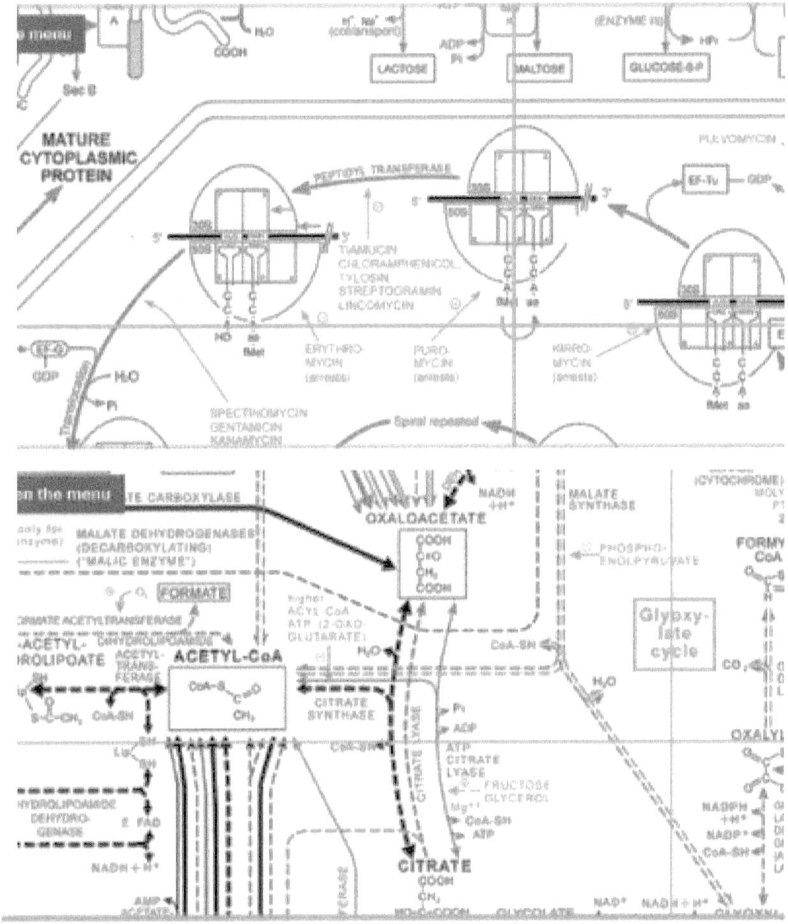

Metabolic Pathways

Mimicking a Neural Network?

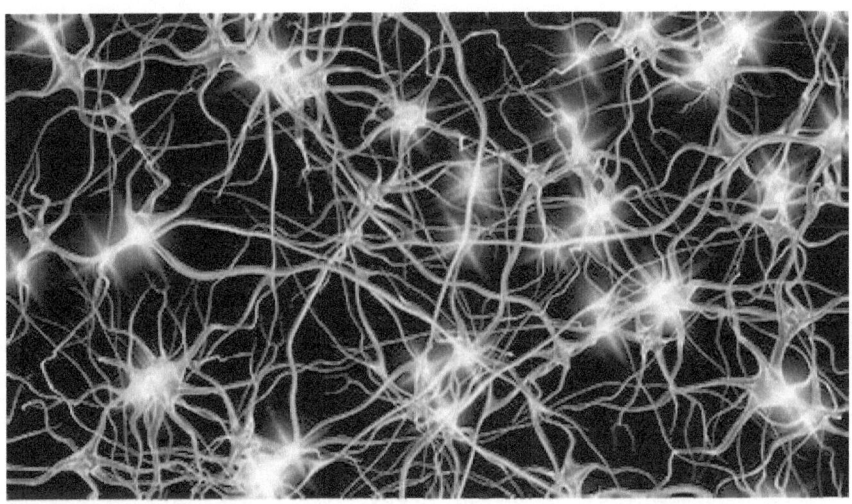

I've come across an interesting article in the Journal of Cyber Security and Information Systems at

https://www.csiac.org/journal-article/enduring-fleeting-future-a-brief-overview-of-current-sentiment-and-emotional-analysis-a-look-forward/

with the interesting -- and just a bit scary -- title "*Enduring, Fleeting, Future: A brief overview of current sentiment and emotional analysis, a look forward.*" This article appears to make the case for researchers seeking out designs in nature in order to design and build "people designed" capabilities of a similar sort. In this case, the problem being targeted is Artificial Intelligence (AI) seeking to recognize and identify particular human emotions of sentiment and emotion, as well as, I suppose, identify one particular face among the many billions of faces throughout the world.

In terms of human emotion, the focus will be on the seven universal emotions identified as: joy, surprise, fear, anger, sadness, disgust, and contempt.

For human sentiment, the following Merriam-Webster Dictionary seems to apply; a) : *an attitude, thought, or judgment prompted by feeling : predilection or b) : a specific view or notion : opinion.*

Pretty scary stuff, and that's perhaps why this article shows up in a journal dealing with Cyber Security.

This article brought to mind an article I wrote a few years back; *"I've grown accustomed to your face"*, about the amazing capability of humans to recognize a face in a crowded room. This capability is further augmented by a similar capacity to recognize particular voices among many. The paper was promoted to "headline" status at the Intelligent design site https://uncommondescent.com/

I have a 50+ year old degree in mathematics, and no knowledge of Artificial Intelligence, but I was able to read the article and get the gist of what was being said. It seems the article can be summed up by the following three pictures:

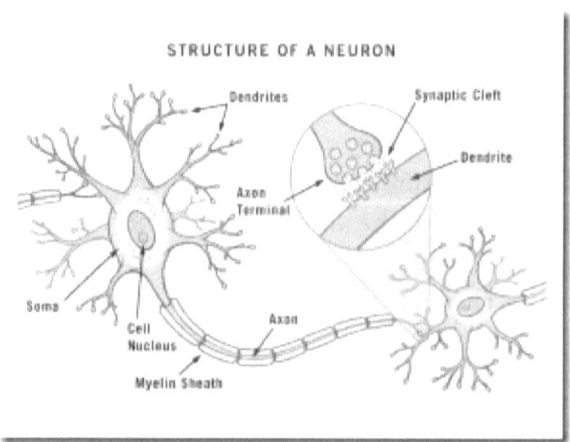

First is a schematic of an actual neuron. There are an estimated 100 billion neurons in a typical healthy human. For an earth's population of some 7.5 billion, that's a lot of neurons that work in order for all those kids to catch a view of mom's dreaded "look."

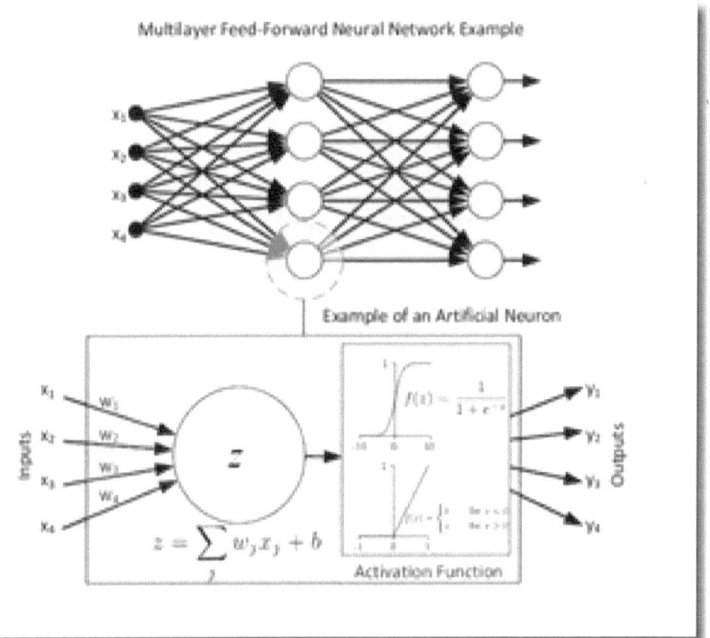

The second picture is a schematic of something called a Multilayer Feed-Forward Neural Network. In the context of the article, this seems to be a human created primitive neuron wannabe.

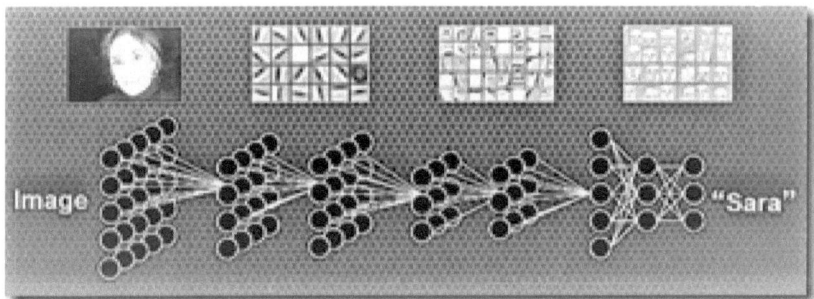

The third picture represents the various stages that take a "convolutional neural network" through the processes of picking out Sara in that crowded room.

The article is filled with technical design jargon, mathematics and whatever. Not once is the term Intelligent Design seen. Likewise, the term evolution does not show up. However, what I see as a layman, very clearly as the underlying design assumption, is an amazing device designed, engineered and manufactured by a mind far beyond any of us mortal humans.

Take another look at that second picture above and for the two mathematical functions, substitute your favorite random number generator. What do you expect to see as output on the right side of the box?

And I would remind you that you don't have to be a PhD scientist to grasp the overall theme being presented here, the scientific notations are indicative of processes that can be grasped by most lay-people.

It would seem Darwin got it wrong.

I praise you because I am fearfully and wonderfully made; your works are wonderful, I know that full well. Psalm 139:14

I've Grown Accustomed to Your Face

The following scenario is familiar to most of us, particularly as we grow older:

We walk into a crowded and noisy room full of mostly strangers and unfamiliar heads bobbing up and down. Then off to the side and slightly behind we hear and recognize a familiar voice ... we turn our head searching for that old friend we know is there, and after a short search ... there she is, head slightly turned away from our view, but recognizable none-the-less. We are surprised and pleased to meet our old friend once more after some number of years and begin renewing the friendship.

The recognition of the voice and face is instinctive and very quick; and we take it for granted with no thoughts of anything unusual other than the mere co-incidence of the meeting.

But behind the scenes in our ears, eyes, nerves, and brains is a marvelous and miraculous process called pattern recognition. A pattern recognition that is able to pick out and recognize individual faces and voices out of the billions of faces and voices surrounding us in the world. So, let's take a brief tour of what's involved in meeting up with our old friend.

The hearing system that most of us have is a partnership between our ears and brain along with the connecting nerves between the two. This stereo audio system is able to sift through the many amplitudes (volumes) presented – the multitudes of widely spread and finely differentiated frequencies – the various timbres, tone color or tone quality, presented by the many voices surrounding us in that room full of strangers. And we are able to pick out that distinctive and familiar voice among the multitudes. And by the way, that same set of ears, in the form of the semi-circular canals, is instrumental in our balance system which keeps us from stumbling around in that crowded room.

And the eyes ... my gosh what a gift ... a gift of obvious design which enables us to stand in awe at the many wonders of our everyday world.

The eyes, as with the ears, are continually involved in a massive process of pattern recognition that allow us to function smoothly within our very busy, active, and dangerous world. Eyes that are quick to warn us of the dangers of that car moving too close to us on the freeway. Eyes that quickly recognize that old friend even in a crowded and busy room.

In our modern technological world, we have analogies to that busy room. Our Navy ships scan the depths of the ocean with sonar. The pulses transmitted from the sonar antenna bounce off; the ocean floor, schools of fish and even the surface of the ocean, returning a bewildering stream of noise that the computers of the sonar must sift through, filter and cluster to present the operators and commanders an array of potential hazards and threats to the fleet. These sophisticated sonar systems require sophisticated computational systems and large amounts of memory storage to accomplish the task in real-time. But most fundamentally they require intelligent designers to create the systems required.

Pattern recognition in the visual world is no less wondrous. When you take a picture of that group at a reunion with a modern state-of-the-art camera, have you noticed the little boxes surrounding the faces? Somehow some very smart scientists and engineers have figured out how to program a computer in your camera to recognize that human faces are part of the picture and visually highlight them for you. And after you take them you can 'tag' the individual faces with names in programs like Facebook. Again, sophisticated computational power and large amounts of memory storage are required for the job. And, as in the case of sonar processing, intelligent designers are necessary to create the systems required.

Pattern recognition is not a trivial task in the engineering world. A snippet taken from a Wikipedia article on "pattern recognition" reads thus:

For a probabilistic pattern recognizer, the problem is instead to estimate the probability of each possible output label given a particular input instance, i.e., to estimate a function of the form

$$p(\text{label}|x, \theta) = f(x; \theta)$$

where the feature vector input is , and the function f is typically parameterized by some parameters .[4] In a discriminative approach to the problem, f is estimated directly. In a generative

approach, however, the inverse probability $p(x|\text{label})$ *is instead estimated and combined with the prior probability*

$p(\text{label}|\theta)$ *using Bayes' rule, as follows:*

$$p(\text{label}|x, \theta) = \frac{p(x|\text{label})p(\text{label}|\theta)}{\sum_{L \in \text{all labels}} p(x|L)p(L|\theta)}.$$

When the labels are continuously distributed (e.g., in regression analysis), the denominator involves integration rather than summation:

$$p(\text{label}|x, \theta) = \frac{p(x|\text{label})p(\text{label}|\theta)}{\int_{L \in \text{all labels}} p(x|L)p(L|\theta) \, dL}.$$

So, I ask you my friends who believe that Darwinian Evolution ... a belief in unguided, unintelligent, and strictly natural processes; is it reasonable and rational that such a process could guide you to that reunion in a crowded room?

And to those of you who denigrate and insult those of us who believe such natural capabilities are the result of an Intelligent Design (ID), I would ask ... which of us is the IDiot?

A Short Conversation with a Real Scientist

I met a medical research scientist on a flight from Philadelphia to San Antonio (I won't disclose his name here). He sat next to me on this flight, and for the first portion of the flight was reading and scanning some sort of technical journal – it looked to me to be a medical journal. He then took out his laptop and for the remainder of the flight was immersed in some fascinating looking web sites, and some animations that I had seen before. I didn't want to disrupt his work/study, but at one point I interrupted him briefly and apologized for looking over his shoulder at what he was viewing.

The kinds of things the scientist was looking at were obviously some micro-biology scenes, and similar to ones I had seen before and are featured in the videos I refer to below by Drew Berry and David Bolinsky.

In seeing these animations previously, I have wondered how they were produced and how accurate were they in depicting the activities of the cell. Were these animations inferred somehow based on these scientists' study of cellular activity, or were they the result of direct observations via some sort of new and exotic microscopic-like instrumentation? My thoughts were that these sorts of cellular and DNA activities were beyond the capabilities of today's instrumentation.

I was anxious to capture a bit of the scientist's time, but I didn't want to interrupt his engagement in his work – he was on his way to a conference and was obviously doing a bit of last-minute preparation.

So, I sat there next to him fidgeting – closing my eyes to rest, but afraid that if I dozed off, I would miss a chance to query him. I opened an eye periodically to see if he had given up on his laptop ... alas, he was diligent in his work until time came when the announcement was made to shut down and put away all electronic equipment and computers ... my time had finally come.

So, I asked the scientist If I could ask a few questions about him and his work, and he obliged, and I was able to squeeze in a few questions in the limited time before the end of the flight. I see from his bio that the scientist is a PhD in Biochemistry, Molecular Biology, and Biophysics ... a very well educated and smart man.

I asked him about the animations I had seen on his computer, and the animations such as shown in the videos below. How are they produced, are they accurate and are they the result of a great deal of inference or are they the result of some type of direct observations? I'll paraphrase his answers as best I can here, and hope that what I write is close to accurate statements of what I believe he was telling me:

The animations are for the most part the result of observations. Today's instrumentation is actually able to look into the cell and see much of what is going on. For example, in the walking machine called Kinesin seen in the videos, we are able to actually see this machine walking on a type of roadway and pulling along a load of cargo manufactured at one place in the cell to another place in the cell. The roadway is actually seen being constructed in front of the machine, and its de-construction is actually seen behind the machine as it moves along. He affirmed that these animations are for the most part accurate representations of the types of activities of the cell and DNA.

These observations are programmed into a computer program, and as new information and observations are obtained, the program is updated. The programs enable researchers to filter the busyness of the cell and isolate the object under study, such as the kinesin. They are also used to open and stimulate further areas of research.

I asked him about the animations showing DNA replication and the error detecting and error correction mechanisms of this machine.

Again, as best I can recall, he said these animations were the result of observations where they can actually see the replication of DNA.

In the case of DNA replication, he referred to this machine as a "brilliant machine." I then asked him if this word – machines – is how they describe the activities of the cell. He said Oh yes, they are machines.

In looking into some of the work his lab is engaged in, here's an excerpt of what I found:

Description of Research.

The goal of our research is to understand the cellular machinery responsible for powering cell movements and shaping the architecture of cells, tissues, and organs. Our discovery-based research focuses on the role of the cytoskeleton, molecular motors, and signaling pathways in powering cell migration, muscle contraction, and the transport of internal cell compartments. The pathways investigated in our laboratory are crucial for several normal and pathological processes, including: cell and tissue development, endocytosis, wound healing, immune response, cardiomyopathies, and metastases of tumors. Most of our current efforts are focused on investigating cytoskeletal motors (myosin, dynein, and kinesin). These remarkable nano-machines use chemical energy stored in our cells (in the form of ATP) to generate mechanical force and motion. Cytoskeletal motors are the engines that power muscle contraction, cell migration, intracellular transport, cell division, and cell shape. We are determining how these motors work at the molecular level, how they are physically connected to the machinery they are powering, how they are regulated, how they interact with other motors and signaling networks, and how their fundamental biophysical parameters impact cell function. We are using a range of biochemical, cell biological, single-molecule, and other biophysical techniques to better understand these proteins in health and disease."

A part of the scientist's research has to do with the cellular machines associated with the muscular machines involved in hearing. This especially interests me because of my own history of Ménière's disease and my loss of hearing in recent years – perhaps I'll dig into this area and learn more about the muscles in my ears.

The scientist also suggested that I Google *"systems biology"* to learn more about the types of research going on in today's modern biology and medical research. I told him that I already had done so and will continue to search out new information.

This very brief visit with this working scientist was quite rewarding in that it confirmed some of my own thinking and layman-level research, as well as filling a few holes in my knowledge – and this from a well-qualified man actively working in the fields of micro-biology.

And as I close, I leave you with a dictionary definition of "brilliant" as the scientist used the word in describing the "brilliant machines" working in our cells:

- Brilliant - adjective
- very bright: flashing with light
- very impressive or successful
- extremely intelligent: much more intelligent than most people

In a later correspondence with the scientist, I was corrected by him because of my putting my preferred definition of Brilliant (3) into his mouth, rather than definition 2 which aligns with his thinking.

"Thank you for your interest in my research. I was quite surprised and pleased to find my work highlighted on your blog. However, I must take this opportunity to point out that you may have misinterpreted my use of the word "Brilliant" in our short conversation. I don't remember the details of our discussion, but I am sure my goal was to express the biological wonder of this motor as "Fantastic" or "Amazing," and not "Brilliant" in terms of ideas or thought.

It is important for me to stress that evolution is the best scientific explanation for the functional diversity of molecular motors. In fact, cytoskeletal motors are a wonderful example of evolutionary tuning to adapt to biological needs. To me, the process of evolution is beautiful in its simplicity and robust adaptability, and the concept of evolution is of the highest scientific merit."

Best wishes.

My response follows:

"Thank you for your kind and reasoned reply. I am somewhat amazed to see your response since this blog is very much small potatoes in terms of distribution and readership – and you had no idea who you were talking to on the flight to Texas.

First of all, I must apologize for my assumption that your use of the word "brilliant" was the same as mine ... your use would more aptly fit the second definition as "very impressive or successful," and I thank you for your clarification and I will update my post to reflect that the definition is mine and not yours."

That said, in part based on my background of close to 40 years in software development of complex real-time and distributed systems, and a layman's keen interest in the 'machines' of life lead me to side with the Intelligent Design view of the marvels of nature, and thus my choice of the definition of 'brilliant' that I used in my post. Should you be curious about the thinking and writings of this laymen, I invite you to click on CATEGORIES on the right side of this blog and select 'Intelligent Design'. I don't suggest however that we enter into any kind of debate – your time and energies are much more valuable in searching out ways in which the many fragilities of the human body can be discovered and treated.

Again, I thank you for your response ... I have perused the Perelman resources you provided me and have found them very fascination.

Were I a bit younger, I would love to pursue a second career in the mysteries of life. However, I have no regrets in investing such a large part of my adult life in software development.

Now I invite you to look at a few videos I have discussed here ... I hope you enjoy them. Do a search on these and you will be rewarded by what you see.

Drew Berry: Animations of unseeable biology | Talk Video | TED.com

Excerpt: And so what I'm about to show you is an accurate representation of the actual DNA replication machine that's occurring right now inside your body, at least 2002 biology. So DNA's entering the production line from the left-hand side, and it hits this collection, these miniature biochemical machines, that are pulling apart the DNA strand and making an exact copy. So DNA comes in and hits this blue, doughnut-shaped structure and it's ripped apart into its two strands. One strand can be copied directly, and you can see these things spooling off to the bottom there. But things aren't so simple for the other strand because it must be copied backwards. So it's thrown out repeatedly in these loops and copied one section at a time, creating two new DNA molecules.

On further review of this amazing animation, I looked at the comments to see what reactions others had, and for further comments and amplifications by Drew Berry. I recommend you do the same, but I include here a particular question and an answer by Drew. I will also list some of the links Mr. Berry points out for further reference and excitement ...

Arthur Brogard

This is the greatest thing. I wonder if we could get more information as to the scientific accuracy of what is portrayed?

The sizes and shapes of the molecules – are they all guaranteed 100% correct? If not when what are the tolerances, just what should we bear in mind when watching?

The behavior of the molecules most especially those astounding 'walking' things – how were they arrived at? We can take it as real that's how they look and how they operate?

Why/how do they operate like that? What causes a leg to swing forward and clamp on? And then why does it release?

And the contents of the cell – all that amazing crowd of structures and proteins, whatever. How real is that? What's the best guess or is no guess needed? Is it 'modified' to better enable us to see what's happening or is it like that, just that congested, just that open?

What's the overall picture? Given the nature of it all I'd imagine thousands, maybe millions, of these processes never come to fruition, in this mad scramble, this ceaseless chaos. So what's the scientific consensus on the state of it all? That a small minority of these processes succeeds or that the vast majority succeed, and we rely heavily on 'braking' mechanisms to stop them? Is the human body in a sort of chemical equilibrium at all times ...?

Those kind of things I'd love to see/hear and conveniently collected up hereabouts for me – I realize it's all available out there somewhere in the chemistry textbooks but......

It is wonderful

I will have these animations playing on computer screens almost permanently in my home.

I'll get wall posters showing stills....

Hi Arthur,

Glad you like the animations and great questions! The molecule models were obtained from Xray crystallography which determines the position of the atoms in each protein. I then created a surface representation of the exterior of each protein to use in the animation. Visit pdb.org to find out more about such data.

The walking molecules (Dynein and Kinesin) were modelled on multiple forms of data. The structures are an assembly of multiple PDB models and the way they move has been a very active area of research for a couple of decades. The leader in this field of molecular research is Dr Ron Vale at UCSF and his work was the primary influence on the development of that part of the animation. Visit his lab's webpage here for more information about what we know about the structure and how the molecular motors move:

https://valelab.ucsf.edu

The molecular world inside the cell is vastly more crowded than I represented it. I made aesthetic and visual communication choices to limit the density and complexity to make this watchable to the audience. To get a better sense of the dense, crowded nature of the molecular world, I highly recommend (!) the work of Dr David Goodsell, his superb illustrations and specifically his illuminating books. His writing style is very accessible to the public and his images are mind blowing in the world they reveal with unrivalled scientific accuracy and detail: David Goodsell's illustrations:

http://mgl.scripps.edu/people/goodsell/illustration/public

Goodsell's books:

http://www.amazon.com/The-Machinery-Life-David-Goodsell/dp/0387849246

Hope this helps answer some of your questions and also gives you much new material to explore and learn from.

Also visit my work homepage for more such animations at WEHI.TV by my small team.

http://www.wehi.tv

Drew

And now some of the links:

https://valelab.ucsf.edu

http://mgl.scripps.edu/people/goodsell/illustration/public

http://www.amazon.com/The-Machinery-Life-David-Goodsell/dp/0387849246

http://www.wehi.tv

http://www.biointeractive.org

https://www.wehi.edu.au/wehi-tv/chromosome-and-kinetochore

David Bolinsky: Visualizing the wonder of a living cell | Talk Video | TED.com

Excerpt: But these machines that power the inside of the cells are really quite amazing, and they really are the basis of all life because all of these machines interact with each other. They pass information to each other. They cause different things to happen inside the cell. And the cell will actually manufacture the parts that it needs on the fly, from information that's brought from the nucleus by molecules that read the genes. No life, from the smallest life to everybody here, would be possible without these little micro-machines. In fact, it would really, in the absence of these machines, have made the attendance here, Chris, really quite sparse.

Excerpt: But it's really quite amazing that these cells, these micro-machines, are aware enough of what the cell needs that they do their bidding. They work together. They make the cell do what it needs to do. And their working together helps our bodies — huge entities that they will never see — function properly. Enjoy the rest of the show. Thank you.

A word about "Deep Time"

The theory of Darwinian Evolution is critically dependent on the idea of "Deep Time." The idea, as expressed by Richard Dawkins as "Climbing Mount Improbable" - with a book by that title - is that given enough time and enough opportunities, anything is possible, indeed probable – thus *'from ooze to us.'*

But putting on my common-sense hat, and observations of a long lifetime, I see that in my own observations at least, things degrade over time, and nothing improves by them just sitting there.

My own body ages and things start wearing out and hurting. I can't see or hear or run like I used to.

I see old cars by the thousands in wrecking yards, rusting away and none of them driving out of the yard as a new and improved model. I don't even have to go to the wrecking yard to see this, I've had many cars in my lifetime, and they all have wound up in those wrecking yards after their years of service is finished.

I go to the mountains and look at the peaks and valleys. I see the mountains crumbling and cliffs collapsing into the valleys below. I see the trees in the forest age and rot – eventually even the ancient Redwoods and Bristlecone Pines. Astronomers and cosmologists tell me the sun will eventually burn out.

I've not seen this idea of time improving things. Oh yes, athletically my body did improve from birth through maybe the twenties, but then Father Time shows up.

But then I am only one person with one lifetime of experience. Perhaps Richard Dawkins, Jerry Coyne, Neil deGrasse Tyson, Carl Sagan and Stephen Hawking have had different experiences.

I'm just thinking ... how would I build that?

or maybe ... maybe it just sort of "poofed" into existence.

Machines everywhere!

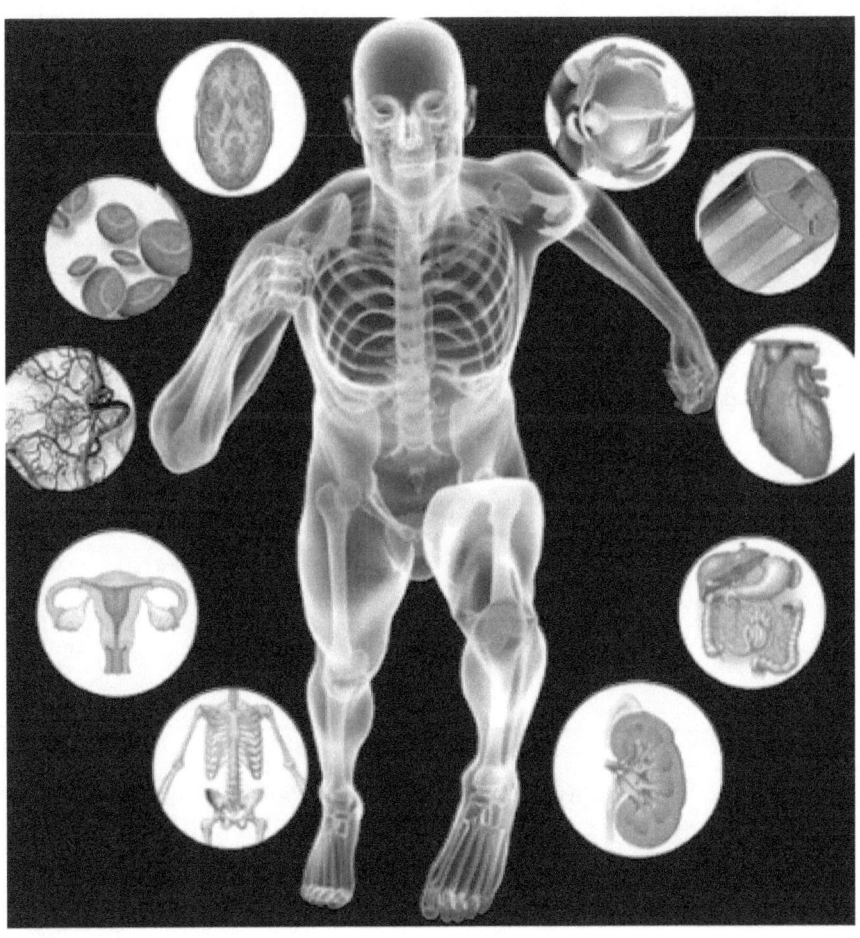

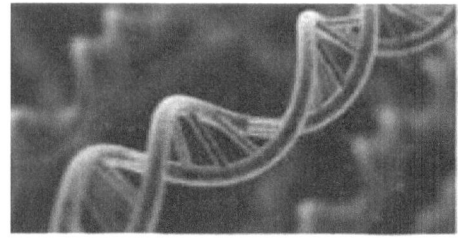
DNA & the cell – a type of blueprint and manufacturing plan.

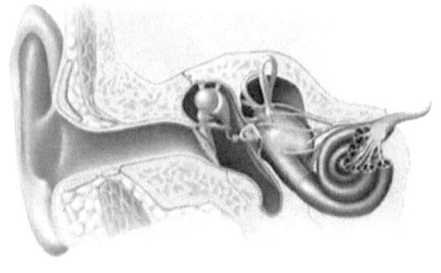
The inner ear – a strapdown inertial reference system. Much like what's used in missiles and aircraft, and also a key machine for balance. It also contains mechanisms for hearing over a wide range of sounds.

The liver – performing over five hundred tasks, each vital for survival.

Flagellum – a cellular outboard motor that moves bacterium from place to place. It's irreducibly complex.

The human eye is capable of detecting a single photon, the smallest unit of light, according to a new study. (Alipasha Vaziri & IMP)

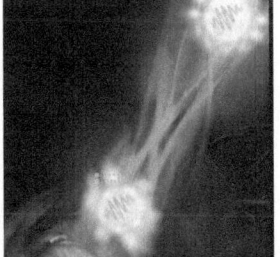

Richard Dawkins: "Any engineer would naturally assume that the photocells would point towards the light, with their wires leading backwards towards the brain. He would laugh at any suggestion that the photocells might point away from the light, with their wires departing on the side nearest the light. Yet this is exactly what happens in all vertebrate retinas. Each photocell is, in effect, wired in backwards, with its wire sticking out on the side nearest the light. The wire has to travel over the surface of the retina, to a point where it dives through a hole in the retina (the so-called "blind spot") to join the optic nerve."

Dawkins explains that the eye "magically transforms from a flat sheet to a cup!"

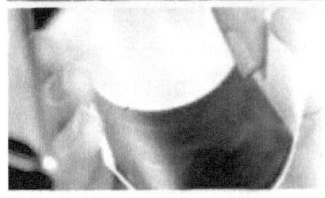

If the hands won't do, then he'll try the polaroid film *"evidence."*

Smartest man in the world??

And this from an article by Casey Luskin, PHD ---

https://intelligentdesign.org/articles/molecular-machines-in-the-cell/

Molecular machines have posed a stark challenge to those who seek to understand them in Darwinian terms as the products of an undirected process. In his 1996 book Darwin's Black Box: The Biochemical Challenge to Evolution, biochemist Michael Behe explained the surprising discovery that life is based upon machines:

Shortly after 1950 science advanced to the point where it could determine the shapes and properties of a few of the molecules that make up living organisms. Slowly, painstakingly, the structures of more and more biological molecules were elucidated, and the way they work inferred from countless experiments. The cumulative results show with piercing clarity that life is based on machines — machines made of molecules! Molecular machines haul cargo from one place in the cell to another along "highways" made of other molecules, while still others act as cables, ropes, and pulleys to hold the cell in shape. Machines turn cellular switches on and off, sometimes killing the cell or causing it to grow. Solar-powered machines capture the energy of photons and store it in chemicals. Electrical machines allow current to flow through nerves. Manufacturing machines build other molecular machines, as well as themselves. Cells swim using machines, copy themselves with machinery, ingest food with machinery. In short, highly sophisticated molecular machines control every cellular process. Thus, the details of life are finely calibrated and the machinery of life enormously complex.

Jargon and Mumbo-Jumbo ... where's the Evidence?

For my thoughts are not your thoughts, neither are your ways my ways, saith the Lord. For as the heavens are higher than the earth, so are my ways higher than your ways, and my thoughts than your thoughts. Isaiah 55:8-9.

A common critique of Intelligent Design is stated as *"Where is the evidence for ID or creation?"*

The evidence is as close to you as your own head, so let me present the evidence from the top down, the top being the human head.

The human head is an amazing collection of instrumentation, computing power and inventive genius that far surpasses any collection of machines brought about by those same heads.

Let's begin.

The array of instruments contained in the head gather in an enormous and varied amount of data allowing for a very active life for the typical human being:

Stereoscopic ears gathering in a wide range of sounds including music of all sorts, conversational speech allowing intimate communication and a gathering of knowledge and inspiration from friends and strangers alike, warnings of danger and much more. Those same stereoscopic ears play a key role in allowing us to know our place in the three-dimensional space we all live in. This is called balance. I know and greatly appreciate this sense of balance because in past years I have suffered from Ménière's disease, a debilitating disease characterized by extreme vertigo and nausea. The inner ear and its gyroscopic characteristics provide this very necessary sense of place and when it gets out of whack all you can do is lie flat on your back and hope it goes away soon. The signals from the inner ear containing balance information are multiplexed along with sound onto nerves that transmit all of this data to the brain which makes final sense of it all.

The eyes are a marvel of design that has been likened to having 12 movies cameras in our eyes. These cameras capture not only color and shapes, but motion of all sorts. The eyes are a major source of data which avails us the limitless literature, art and science of the world we live in. Education, for the most part, enters our being through our eyes. The eyes are also a major contributor to what I call the "sixth sense"; that of aesthetics . . . that sense of appreciation for the beauty and wonder of the world around us:

- The beautiful panorama of a distant mountain range.
- The peaceful beauty of an ambling, bubbling stream in a meadow of flowers.
- The majesty of a moonless sky.
- The sensuous beauty of a loved one.

The nose, though of lesser notice than sight and sound, provides us with the pleasures of the smell of a good meal, the scent of an infant child, the romance of a splash of perfume placed just right, springtime flowers and more. The nose also provides warnings of things we might not want to eat.

Then there's taste. The taste of a well-cooked meal, the supple texture and taste of a bucket of mussels or clams. Coupled with smell, taste is the source of much pleasure in the life of the typical human being.

Now we move down the body to see how this instrumentation and the resultant data are translated into meaningful actions and tangible artifacts.

These instruments (remember, I'm talking about these things as being designed) transmit their received data to the brain through a sophisticated system of nerves. At the brain, this vast, complex, and ever-changing array of data is synthesized into a coherent audio/visual/mental picture of our surroundings. Indeed, the brain gathers and stores this data, and allows us to transfer this data in modified form to those around us in the form of literature, art, conversation, music, musings, and all manner of creativity, including science and engineering.

The hands at the end of our arms cradle quill and ink, pen and ink, keyboard, hammer, chisel, saw and lathe … to craft the great literature, music and architecture of civilization. The mind commands the hands to paint the great art we see and enjoy in museums around the world; same with the great sculptures that capture our gaze.

We marvel at the great cathedrals and monuments to great and memorable people. We live in comfortable homes designed by the creative minds and hands of architects, and built by the hands of great craftsmen, plumbers, and electricians.

We travel around the world on the great railroads, ships, and airlines, designed and built by the minds and hands of designers and engineers and built by hundreds of skilled hands.

Now to complete the picture we travel to the legs and feet, and to the world of sport.

We marvel at the skill of baseball players as they complete the intricate double play. The dribbling, ball handling and shooting skills of a Michael Jordan captivate a cheering audience. Thousands' marvel and cheer at race tracks around the world as skilled drivers navigate complex race courses. The athletic skill and training of airshow pilots attracts multitudes.

All these human talents and skills come from an intelligently designed body and mind focused on the pursuit of perfection.

At the scientific and investigative end of this spectrum we find scientists pondering over the marvelous world of DNA, a world that shouts loudly; ***information ... design ... intelligence***. A world where living cells are programmed to become ... bone, skin, brain, muscle ... and at the exactly appropriate time for development of a fully functional human being.

And we see those scientists discover and explore the extreme finely tuned universe and earth which are required for life to exist at all. So, you see ... the evidence is there for you to see, hear, smell, taste, touch . . . and enjoy; only a touch away with a finger to the cheek.

Richard Dawkins says, "Biology is the study of complicated things that give the appearance of having been designed for a purpose."

I say "If it looks like a (designed) duck, walks like a (designed) duck and quacks like a (designed) duck. There is a very good chance it is a designed duck." And I don't have a prestigious seat at a prestigious university.

Here is the Evidence

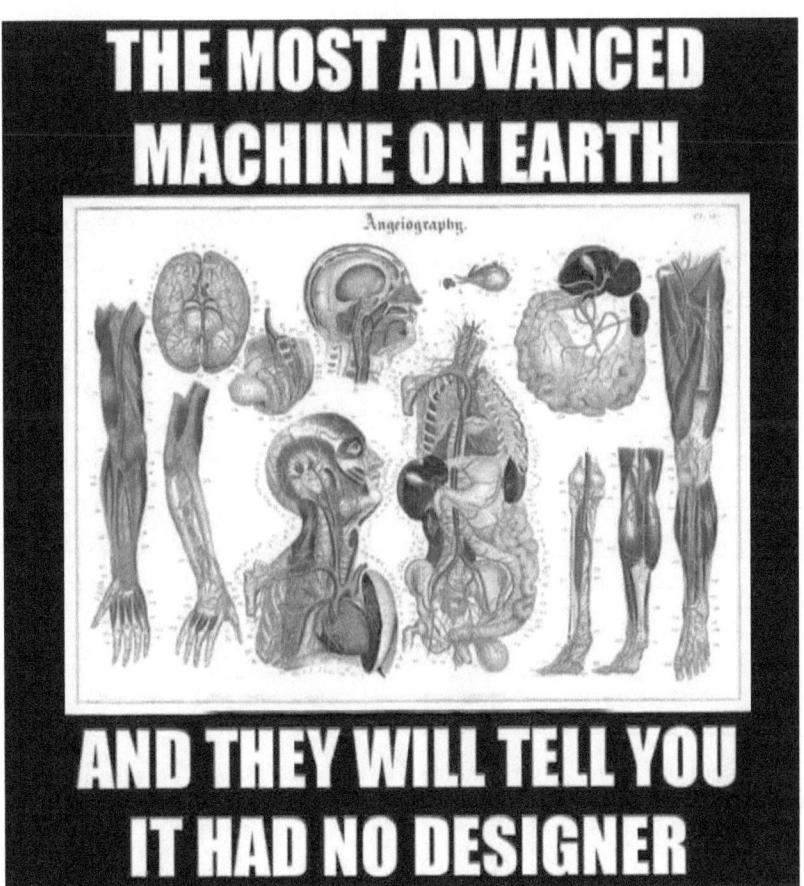

Massively Complex Synchronicity – Part 1

This and the following two sections are from essays I wrote back in 2012 in my attempts to wrap my mind around the vast complexity I see all around me.

In the beginning God created the heaven and the earth. Genisis 1:1

The scope of that simple statement is breathtaking. The heavens in all their glory! Multiple upon multiple upon multiple billions of stars and galaxies stretching out as far as we can see with the Hubble and Webb Telescopes to an astounding 31,000,000 light years away and the Whirlpool Galaxy. Now a light year is what we use to measure such distances, and it is not your ordinary household tape measure. A light Year is the distance that light travels at 186,000 miles/second in a year; or 6 trillion miles. So we have something that we can see and marvel over that is 31,000,000 x 6,000,000,000,000 miles away. Do the math, its a long, long, long way away. It's even further away than the furthest Starbucks Coffee shop in my town when I run out of coffee early in the morning ... now that's far!

"Who else has held the oceans in his hand? Who has measured off the heavens with his fingers?..." Isaiah 40:12

"He determines the number of the stars and calls them each by name." Psalm 147:4

At the other end and close to home we have the really– really– really– really small. And it is just as incredible as the really–really– really– really large! Such as the DNA molecules which define life. And those really-really-really-really small things in all of us; the DNA, the chromosomes, the cells, the synapsis in our brains, all work together so we can think about and invent things like the Hubble Space Telescope so we can search out the mysteries of the really-really- really- really big things like the Whirlpool Galaxy!

The Laminin cell.

And just two of those really small things, a single cell from your mother and a single cell from your father come together, each with half of their own chromosome, to form an entirely new and distinct person; you!

This new cell from your parent's cells is somehow smart enough to split and divide and somehow know how and when to send some to become bone, some to become skin, some to become heart, some to become eyes, some to become feet and some to become hair. And would you believe it, the master designer of it all claims:

> "As for you even the hairs of tour head have all been counted." And as you go through life that same designer says "Fear not, for I have redeemed you; I have called you by your name; You are Mine."
> Isaiah 43:1

So, He not only knows and names the really, really, really, really large, and the hairs on each of our heads, but He knows each of us by name. Wow! Can you imagine going to a Super Bowl game and watching someone going row by row, seat by seat and greeting each and every one by name, and not missing even one?

> "Who else has held the oceans in his hand? Who has measured off the heavens with his fingers?..." Isaiah 40:12

So let's go back to those two images we saw above; first the Whirlpool Galaxy. Take a look at the very center of that galaxy. And then watch [this clip](#) (search for 'Louie Giglio Indescribable'.)

Then take a look at the laminin cell and watch this clip (search for Louie Giglio Laminin.)

.

But that's not the end of it. Scientists now tell us that as much as we know about the universe, what we know is only about 4% of what they think is really there. The rest, the unknown 96 percent, is made up of Dark Matter and Dark Energy of which nothing at all is known. And just the other day I found an article talking about Dark Force, an unknown force that is apparently pulling large portions of our universe to somewhere else.

My thoughts on these things will have to wait for a Part 2 of this series, so stay tuned. In the meantime, enjoy the ideas I have presented, and take time to explore my new found spiritual hero Louie Giglio.

Don Johnson – May 2012

Massively Complex Synchronicity – Part 2

The Origins of the Universe ... Simple or Complex: The Problem of "Massively Complex Synchronicity"

For thus says the LORD that created the heavens; God himself that formed the earth and made it; he has established it, he created it not in vain, he formed it to be inhabited: I am the LORD; and there is none else.   Isaiah 45:18

My <u>earlier post</u> on the subject focused on the immensity of the universe and the amazing complexity and functionality at the micro-biological level. In this essay I would like to focus on a concept I've labeled as *"Massively Complex Synchronicity."*

My purpose in this essay is to summarize evidence of design across the spectrum from the micro to the macro level and thus encourage you, the reader, to investigate further. After all, if there is design, then there must be a designer, and if there is a designer, then what is his claim on your life? Much like the <u>researchers looking for the "God particle"</u>, my hope is to show you something similar to finding the fossilized imprint of a dinosaur: "You see the footprints and the shadow of the object, but you don't actually see it." My motivation is to cause you to seek the God of the Bible, and His Son Jesus Christ.

(See my Bio below at the end of this essay:)

By synchronicity, I mean the relationship between two or more things resulting in something mutually beneficial, or in some cases a new thing arising from the synchronicity.

In the debate between Darwinian Evolution and Creation and Intelligent Design, the concept of *"irreducible complexity"* comes into play where it is claimed that some things are sufficiently complex such that if any component is removed from the system, then the whole thing ceases to exist in it's accepted form. The theory of <u>intelligent</u>

design holds that certain features of the universe and of living things are best explained by an intelligent cause, not an undirected process such as natural selection. The mousetrap is often cited as a simple analogy where if any of it's component parts is removed it ceases to be a mouse trap. Biological systems cited as having irreducible complexity characteristics include the light-sensing mechanism in eyes, the human blood-clotting system, and the bacterial flagellum.

Critics of the irreducible complexity stance counter with various experiments/studies which they claim falsify the concept of irreducible complexity.

Although the debate is very interesting, my intent in this essay is not to contribute to this debate, but rather to elevate the discussion to a higher level and illuminate the massive complexity we observe in all systems whether they be cosmological, biological or micro-biological in nature. My premise is that in this complexity there exists a majestic synchronicity that is necessary for life to exist at all.

The ID proponents will continue to point out new discoveries that point to an intelligent designer, and the Darwinists will continue to counter point by point, often with pointed rebuttals of the particulars involved, but most often with a shotgun approach claiming Darwinian evolution as settled science ... *end of discussion!*

So, lets move beyond the point by point example/counter-example and look at an admittedly very partial picture of the Massively Complex Synchronicity present all around us; a synchronicity which is necessary for our very existence as an individual person on this very individual planet in this very large universe.

Our human body is the first stop on our trip through the synchronicity of the universe, and we begin at the cellular level with a quote from Dr. Michael Behe in his book Darwin's Black Box.:

The entire cell can be viewed as a factory that contains an elaborate network of interlocking assembly lines, each of which is composed of a set of large protein machines. . . . Why do we call the large protein assemblies that underlie cell function protein machines? Precisely because, like machines invented by humans to deal efficiently with the macroscopic world, these protein assemblies contain highly coordinated moving parts.

Now take a look at some of the Molecular Machines that Scientists Have Argued Show Irreducible Complexity. For brevity's sake I have listed only the names of the machines but have left a few descriptions to whet your appetite (Note: Not all of the items listed exist in human cells, nor do they exist in all cells). My thanks to Casey Luskin of Discovery Institute for this list.

1. **Bacterial Flagellum:** The flagellum is a rotary motor in bacteria that drives a propeller to spin, much like an outboard motor, powered by ion flow to drive rotary motion. Capable of spinning up to 100,000 rpm,[13] one paper in Trends in Microbiology called the flagellum "an exquisitely engineered chemi-osmotic nanomachine; nature's most powerful rotary motor, harnessing a transmembrane ion-motive force to drive a filamentous propeller."[14] Due to its motor-like structure and internal parts, one molecular biologist wrote in the journal Cell, "[m]ore so than other motors, the flagellum resembles a machine designed by a human."[15] Genetic knockout experiments have shown that the E. coli flagellum is irreducibly complex with respect to its approximately 35 genes.[16] Despite the fact that this is one of the best studied molecular machines, a 2006 review article in Nature Reviews Microbiology admitted that "the flagellar research community has scarcely begun to consider how these systems have evolved."[17]

2. **Eukaryotic Cilium:**

4. **Blood clotting cascade:**

5. **Ribosome:**

6. Antibodies and the Adaptive Immune System: Antibodies are "the 'fingers' of the blind immune system—they allow it to distinguish a foreign invader from the body itself."[30] But the processes that generate antibodies require a suite of molecular machines.[31] Lymphocyte cells in the blood produce antibodies by mixing and match portions of special genes to produce over 100,000,000 varieties of antibodies.[32] This "adaptive immune system" allows the body to tag and destroy most invaders. Michael Behe argues that this system is irreducibly complex because many components must be present for it to function: "A large repertoire of antibodies won't do much good if there is no system to kill invaders. A system to kill invaders won't do much good if there's no way to identify them. At each step we are stopped not only by local system problems, but also by requirements of the integrated system."[33]

II. Additional Molecular Machines

7. Spliceosome:

8. F_oF_I ATP Synthase:

9. Bacteriorhdopsin: Bacteriorhodopsin "is a compact molecular machine" uses that sunlight energy to pump protons across a membrane.[39] Embedded in the cell membrane, it consists of seven helical structures that span the membrane. It also contains retinal, a molecule which changes shape after absorbing light. Photons captured by retinal are forced through the seven helices to the outside of the membrane.[40] When protons flow back through the membrane, ATP is formed.

10. Myosin:

11. Kinesin Motor:

12. Tim/Tom Systems:

13. Calcium Pump: The calcium pump is an "amazing machine with several moving parts" that transfers calcium ions across the cell

membrane. It is a machine that uses a 4-step cycle during the pump process.[47]

14. Cytochrome C Oxidase:

15. Proteosome: The proteosome is a large molecular machine whose parts must be must be carefully assembled in a particular order. For example, the 26S proteosome has 33 distinct subunits which enable it to perform its function to degrade and destroy proteins that have been misfolded in the cell or otherwise tagged for destruction.[50] One paper suggested that a particular eukaryotic proteasome "is the core complex of an energy-dependent protein degradation machinery that equals the protein synthesis machinery in its complexity."[51]

16. Cohesin:

17. Condensin:

18. ClpX:

19. Immunological Synapse: .[56]

20. Glideosome:

21. Kex2:

22. Hsp70:

23. Hsp60:

24. Protein Kinase C: Protein Kinase C is a molecular machine that is activated by certain calcium and diacylglycerol signals in the cell. It thus acts as an interpreter of electrical signals, as one paper in *Cell* wrote: "This decoding mechanism may explain how cPKC isoforms can selectively control different cellular processes by relying on selective patterns of calcium and diacylglycerol signals."[62]

25. SecYEG PreProtein Translocation Channel:

26. Hemoglobin:

27. T4 DNA Packaging Motor:

28. Smc5/Smc6:

29. Cytplasmic Dynein: Cytplasmic dynein is a machine involved with cargo transport and movement cell that functions like a motor with a "power stroke."[70] In particular, it transports nuclei in fungi and neurons in mammalian brains.[71]

30. Mitotic Spindle Machine:

31. DNA Polymerase:

32. RNA Polymerase: Like its DNA polymerase counterpart, the function of the RNA polymerase is to create a messenger RNA strand from a DNA template strand. Called "a huge factory with many moving parts,"[78] it is a "directional machine and, indeed, as a molecular motor" where it functions "as a dynamic, fluctuating, molecular motor capable of producing force and torque."[79]

33. Kinetochore:

34. MRX Complex:

35. Apoptosome / Caspase:

36. Type III Secretory System:

37. Type II Secretion Apparatus:

38. Helicase/Topoisomerase Machine: The helicase and topoisomerase machines work together to properly unwrap or unzip DNA prior to transcription of DNA into mRNA or DNA replication.[90] Topoisomerase performs this function by cutting one DNA strand and then holding on to the other while the cut strand unwinds.[91]

39. RNA degradasome:

40. **Photosynthetic system:** The processes that plants use to convert light into chemical energy a type of molecular machines.[95] For example, photosystem 1 contains over three dozen proteins and many chlorophyll and other molecules which convert light energy into useful energy in the cell. "Antenna" molecules help increase the amount of light absorbed.[96] Many complex molecules are necessary for this pathway to function properly.

I will give thanks unto thee; for I am fearfully and wonderfully made: Wonderful are thy works; And that my soul knoweth right well. Psalm 139:14

So you see we have massive complexity at the molecular level, and science being what it is, we can be reasonably assured that the above list is nowhere near complete. Sure, other scientists can chip away at some individual items on the list and "debunk" claims of irreducible complexity, but how can you escape the overall imprint of complexity and the underlying argument for design?

These sophisticated molecular machines, of course, reside in the various organs making up the human body, and there are many such machines in each organ, and these organs are themselves machines and marvels of specialty and functionality.

But as marvelous as these organs are, they do not exist alone, but in synchronicity with others that make up a human body. And these organs do indeed represent irreducible complexity. A heart that fails kills many a person, as does a failed kidney, liver, pancreas, stomach, brain, spleen or lung, although some of these organs are redundantly created in pairs and a body can function with only one of the pairs.

Redundancy of body parts and organs is another example of being "wonderfully made." Redundancy in our eyes, ears and the inner ears give us a keen sense of place in our three dimensional world that add greatly to our human experience in a "wonderfully made" universe, though not necessary.

So now we have a very complex human body made up of millions of machines at the molecular level, and dozens at the organ level. Each machine has a specific purpose and thus a specified design to carry out that purpose.

But this body does not live in isolation, but rather in massive synchronicity with other such assemblages. At the very basic level, the body must be able to replicate itself in order that the species is able to live beyond one generation; this of course is the mating up of one very specific "male" cell with a correspondingly specific "female" cell resulting the creation of a new and unique body called a child. And notice that this union is accompanied along with a phenomenon known as "pleasure"; a phenomenon which Darwinists are hard pressed to explain or categorize into the materialist viewpoint in which their theory of necessity resides. So we have here a sort of "Irreducible Complexity", not within the same cell or even among organs, but a complexity requiring an equal contribution from two independent and completely functional persons; a mother and a father.

From here on out I am going to put the camera of life into a time lapse mode and take glimpses of the massively complex synchronicity required to sustain the three lives I've just pictured.

 The three lives; mother, father and child (a family) must have some sort of life support system available to take in the continual stream of new energy required to replenish the energy expended just simply to stay alive. This is called food and takes on the basic form as described above; a complex assemblage of cells, machines and organs making up the vast selection of fruits, vegetables, fish and animals that make up our daily diet. Again, the strange, but not quite material phenomenon of pleasure comes into play as we enjoy the many flavors and textures of the daily diet; mush or gruel would seem to suffice here, but yet we have much more.

All of these life forms required for human existence of course require an exquisitely balanced eco system or it is all for naught;

An atmosphere with just the right sort of mixture of gasses required for the plants and animals to breath and live.

A weather system sufficient to distribute water across a very large planet, and to maintain a livable habitat.

A series of oceans containing an abundance of living creatures usable for food; a mechanism for moderating the climate around the world; a system for distributing warming currents to all area of the earth; ....

A system of fresh water rivers and lakes containing all manner of food stuffs; a system for replenishing the oceans of the world; a delightful place to spend a morning, afternoon, day or longer (there's that pesky emotional thing again that won't be explained away by any kind of "Unified Theory of Everything . Sorry Steven Hawking!")

A system of magnetic bands which protect the fragile earth life from the ravages of solar radiation.

. . .

So that pretty much sums up (but not completely of course) the massively complex synchronicity required for the creation and sustainment of life on earth, and we can now zoom out to [the solar system and universe](#) that supports all of this and the laws and physical constants that undergirds the whole thing;

Underlying what we observe in nature are an array of "constants" which describe natural phenomena and behaviors such as gravity, vacuums, speed of light and many others. I include lists of such constants here not for you to read and study in detail, but to show you the scope and number of such constants. Most likely you are not, nor am I, experts in these fields of science associated with these constants, but I encourage you to peek at a few of them just to get a feel, and perhaps ask yourself questions such as; "what would be the impact if one or more of these constants changed or went missing?" Others have researched such questions, and I refer you here to _Design in Nature: The Anthropic Principle_ by Donald B. DeYoung, Ph.D.

>Constants in the category" Universal constants "...

>Constants in the category" Electromagnetic constants "...

>Constants in the category" Atomic and nuclear constants "...

>Constants in the category" Physico-chemical constants "...

I hope I have created a clear enough picture that has kindled some interest and curiosity in a topic of continuing and universal interest and importance, though not a topic at your average dinner party.

And finally, consider this: scientists can be a strange breed. On the one hand we have those who claim we know only 4% of all there is in the universe and are ignorant of Dark Matter, Dark Energy and Dark Force comprising the remaining 96%. On the other hand are those such as Sagan, Richard Dawkins and Stephen Hawking, who assert with 100% certainty that there is no god and no creation. I suggest that while we treat science and scientists with the respect due them, we should be wary and not buy into their every preaching with little or no skepticism.

The fool says in his heart, "There is no God."   Psalm 14:1

My Bio:

I am not a cosmologist, a paleontologist, an astronomer, a physicist, or a micro-biologist or a college professor.

What I have been by trade and profession for close to 40 years is a software developer ... a computer programmer. I've been directly involved in writing software for some very complex systems; a computer controlled automated warehouse system inputting and outputting orders and transferring products via conveyors and stacker cranes. For close to 30 years, I was involved in the programming of a system that tracks, records and displays the maneuvers and activities of up to 100 high performance fighter aircraft.

So I have developed a keen sense of and appreciation for complexity, a sense gained from many years of getting my hands dirty on a daily basis and experiencing the frustrations of debugging less that perfect systems while trying to perfect them. I have also developed an appreciation for the fact that complex systems require intelligence; the architects, designers and builders of systems ... they do not appear from nowhere contrary to what Darwinists would have me to believe.

I have also studied the debate over the years and have read much on Darwinism, Creation Science and Intelligent Design having come from a Darwinian background myself.

Massively Complex Synchronicity – Part 3

> *I will give thanks unto thee; for I am fearfully and wonderfully made: Wonderful are thy works; And that my soul knoweth right well.*   Psalm 139:14

Just as more capable and powerful telescopes are showing us more of the universe and exposing billions upon billions of new things in the heavens above us, new research at the very small end of the spectrum is showing us more of what life is all about, in particular, human life.

Just yesterday as I was perusing the web I ran across some very spectacular, fascinating and truly exciting research having to do with human genomes (i.e. our DNA).

I begin with some links describing this research:

From Discovery Institute:

Junk No More: ENCODE Project Nature Paper Finds "Biochemical Functions for 80% of the Genome"

From NBC News:

New DNA project shows us living beyond our genes

From the New York Times:

Bits of Mystery DNA, Far From 'Junk,' Play Crucial Role

From the U.S. Department of Health and Human Services – National Institute of Health (NIH) :

ENCODE data describes function of human genome

I chose these reports because they represent a spectrum of thought regarding the nature and origins of life: Discovery Institute takes the view that an intelligent designer is most often the best explanation for the complexity and function of systems such as the human genome. NBC News and the New York Times are main stream news organizations that will typically fall on the side of evolution in explaining such systems, and are often harsh critics of Intelligent Design. NIH is the organization that funded and managed the research projects involved and is included as the truth source.

A couple of snippets from the New York Times:

"The human genome is packed with at least four million gene switches that reside in bits of DNA that once were dismissed as 'junk' but that turn out to play critical roles in controlling how cells, organs and other tissues behave. The discovery, considered a major medical and scientific breakthrough, has enormous implications for human health because many complex diseases appear to be caused by tiny changes in hundreds of gene switches. ... Human DNA is 'a lot more active than we expected, and there are a lot more things happening than we expected,' said Ewan Birney of the European Molecular Biology Laboratory-European Bioinformatics Institute, a lead researcher on the project." (Gina Kolata, "Bits of Mystery DNA, Far From 'Junk,' Play Crucial Role," New York Times (September 5, 2012))

And:

"There also is a sort of DNA wiring system that is almost inconceivably intricate.

"It is like opening a wiring closet and seeing a hairball of wires," said Mark Gerstein, an Encode researcher from Yale. "We tried to unravel this hairball and make it interpretable."

There is another sort of hairball as well: the complex three-dimensional structure of DNA. Human DNA is such a long strand — about 10 feet of DNA stuffed into a microscopic nucleus of a cell — that it fits only because it is tightly wound and coiled around itself. When they looked at the three-dimensional structure — the hairball — Encode researchers discovered that small segments of dark-matter DNA are often quite close to genes they control.

In the past, when they analyzed only the uncoiled length of DNA, those controlling regions appeared to be far from the genes they affect."

And, a couple of snippets from the Discovery Institute:

"Far from consisting mainly of junk that provides evidence against intelligent design, our genome is increasingly revealing itself to be a multidimensional, integrated system in which non-protein-coding DNA performs a wide variety of functions. If anything, it provides evidence for intelligent design. Even apart from possible implications for intelligent design, however, the demise of the myth of junk DNA promises to stimulate more research into the mysteries of the genome. These are exciting times for scientists willing to follow the evidence wherever it leads."

(Jonathan Wells, The Myth of Junk DNA, pp. 9-10 (Discovery Institute Press, 2011).)

I see this study and its stunning report as meshing very nicely into what I have previously called Massively Complex Synchronicity.

It would seem to me that it has become immensely more difficult to explain life and biology with evolutionary "just so" stories. Is it not time to give credence to those who attribute such magnificent designs to an intelligent designer?

Is it not time to seriously consider that the world view beginning with the words "*In the beginning God created the heavens and the earth*" is indeed the correct one?

Don Johnson – September 2012

The Necessity for Simultaneity (i.e. The Creation Week)

All of us living in this world, this universe, no matter where or when - minute-by-minute, hour-by-hour, day-by-day, year-by-year experience the same life phenomena. Lifetimes come and lifetimes go. Life requires that all our surrounding environment, the air we breathe, the food we eat, the water we drink, the animals we eat, the very ground we stand on – all of these must be continually and simultaneously available to us, lest we die.

The evolution story of deep time with plants and animals developing step by precarious step over eons of time is fraught with problems. At each evolutionary step, each kind of plant, each kind of fish, each kind of bird, each kind of mammal must have reproductive capability, this is fundamental. Plant life of some sort is presumably the first life to have evolved, thus from the very first, seeds of some sort must have been available containing the blueprint for following generations. If this evolutionary path of plant life occurred independent of the *"deep time"* evolutionary path of animals, then at what point did the paths converge thus providing the balance of nature we see in our world with the vast variety of foods?

Within our own bodies we have a multitude of codependent organ systems, and we see that a simultaneous, mature, functional completion is necessary. Stereoscopic vision and three-dimensional hearing, along with balance and spatial awareness mechanisms are vital for day-to-day survival. The long eons postulated for developing these traits is a picture of continual disaster negating the possibility of producing the end product of a fully functional human body such as a Michal Jordan.

The Biblical story of creation in contrast, is a model that works to produce the world we live in and experience. The first man, (Adam), and the first woman, (Eve) were created fully formed and mature, including the maturity of their sexual organs. Food was immediately available in the form of various kinds of plants. Breathable air was immediately available; water was there to enjoy.

In shorthand form, Genisis describes this concept of **Massively Complex Synchronicity.**

Your Designed Body: truth and rejection

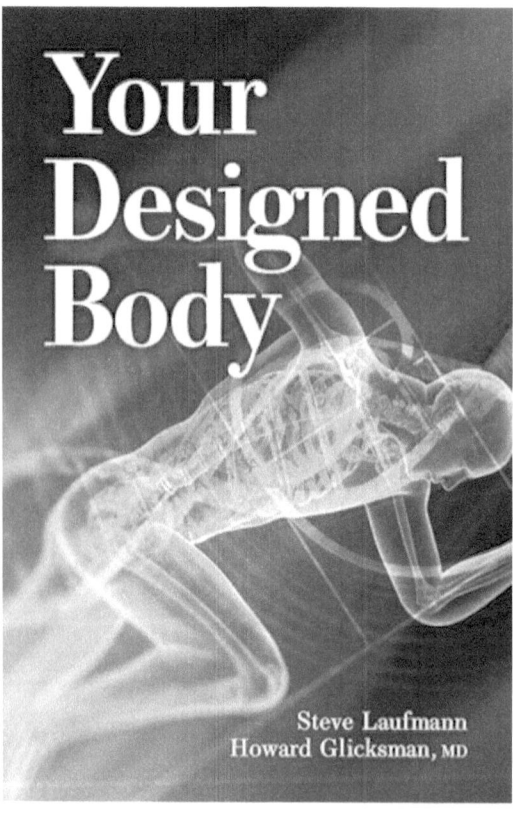

This book from Discovery Institute is an amazing examination of the human body from the perspectives of medicine and systems engineering.

The authors dive deep into all aspects of the human body from the cellular level, each organic system, and the complete functional body, a system of systems.

The emphasis is on the scientific and medical facts associated with each of the body systems, allowing readers to make reasoned and evidence-based judgments on the idea of design in life.

The authors, a medical doctor and a systems engineer, are not theologians, and their work is presented in an objective, scientific and factual manner with minimal theology.

Part II: Evolution, Intelligent Design, Creation/Faith

I've just read an article from Casey Luskin of the Discovery Institute critiquing Eugenie Scott, former director of the National Center for Science Education, on the conflating of Intelligent Design and Biblical Creation Science.

I have had connections with both individuals.

- Eugenie Scott Lecture Resurrects, Spreads Misinformation on Intelligent Design

https://youtu.be/iZhO7_GhDH0 Dr. Scott's presentation

https://evolutionnews.org/2022/01/eugenie-scott-lecture-resurrects-spreads-misinformation-on-intelligent-design/

I've met Luskin several times and have heard him speak at conferences and have read many of his articles. Casey graciously critiqued several of my ID articles a few years ago.

My connection to Eugenie Scott was as an early contributor to the NCSE blog when it went live for public comment a number of years ago. My contributions were not well received, to understate it, often nasty and personal. My comments most often were in rebuttal to articles written in support of NCSE's position of support for Darwinian Evolution in schools. My rebuttals often contained information and views from scientists and thinkers who differed from the orthodox views of evolutionary "science."

Having come to Christian faith from a long period of Atheism, I see ID, not in opposition or conflation to so-called Creation Science, but rather as an off-ramp from an atheistic/evolutionary world view.

My story, and my journey to a Biblical world view had several huge road signs leading to that off-ramp. Creation Science was one of those road signs that alerted me to the hazards ahead had I remained on the freeway of a materialistic world view.

Another huge road sign was prophetic Biblical prophecy, undoubtedly the most prominent of my personal road signs. I was (and still am) amazed in reading about the many, many predictive prophecies of Israel and of Jesus Christ. Prophecies that have uncannily come to fruition, and many clearly seen in fulfillment in today's world. The very existence of modern Israel being one of the most prominent fulfillments.

I began learning of these prophesies from the popular writers of the day such as Hal Lindsey. Early on in my faith journey, I found myself locked away in my bathroom with my stolen Gideons Bible and a copy of Lindsey's book The Late Great Planet Earth (and others), validating Lindsey against that stolen Gideons Bible.

So, I welcome both camps as viable contributors to the quest for understanding of this complex thing called The Universe, and all within it, including you and me.

Part III: Why Does All This Matter?

All this matters because this controversy is a type of censorship. Censorship which affects all of us in some way or another. It affects some in a direct way by threatening careers and livelihoods as the case of Richard Sternberg and the case of Dr. Eric Hedin at Ball State University. It affects young, inquisitive and growing minds by denying exposure to significant and legitimate questions and avenues of inquiry.

It matters because this forced acceptance of only the viewpoint of Darwinian evolution is cited by many young people as a primary reason for leaving the church, and for the loss of faith.

The following is an excerpt from a Pew Research Center survey:

When asked why they didn't believe, many said their views about God had "evolved" and some reported having a "crisis of faith." Their specific explanations included the following statements:

"Learning about evolution when I went away to college"

"Religion is the opiate of the people"

"Rational thought makes religion go out the window"

"Lack of any sort of scientific or specific evidence of a creator"

"I just realized somewhere along the line that I didn't really believe it"

"I'm doing a lot more learning, studying and kind of making decisions myself rather than listening to someone else."

The data from this 2016 study may explain why ex-Christians "question a lot of religious teaching," as reported in the 2018 study. The teaching they question seems to be about the existence of God, and this is consistent with the explanations offered by ex-Christians in a variety of other recent studies. When Christians walk away from the faith, more often than not, it's due to some form of intellectual skepticism. Ex-Christians often describe religious beliefs as innately blind or unreasonable.

The full report can be read at:

http://www.pewresearch.org/fact-tank/2016/08/24/why-americas-nones-left-religion-behind/

A prime motivation for writing this, my book, is to counter such thinking, and to counter this thinking from the point of view and experience of an interested layman. I hope to present the idea that a personal common-sense, experiential examination of the issues and the evidence can lead minds to consider alternatives to the orthodox materialist view of reality.

That controversy over evolution and intelligent design has potential career ending consequences is fully illustrated by the experience of Dr. Richard Sternberg and further illustrated by the case of Dr. Eric Hedin and his position at Ball State University.

The Case of Richard Sternberg

The case of Dr. Richard Sternberg dramatically highlights the professional dangers that come with following the evidence ... if that trail leads one to a conclusion of Intelligent Design.

Read the account as documented by Sternberg:

> "In 2004, in my capacity as editor of The Proceedings of the Biological Society of Washington, I authorized "The Origin of Biological Information and the Higher Taxonomic Categories" by Dr. Stephen Meyer to be published in the journal after passing peer-review. Because Dr. Meyer's article presented scientific evidence for intelligent design in biology, I faced retaliation, defamation, harassment, and a hostile work environment at the Smithsonian's National Museum of Natural History that was designed to force me out as a Research Associate there. These actions were taken by federal government employees acting in concert with an outside advocacy group, the National Center for Science Education. Efforts were also made to get me fired from my job as a staff scientist at the National Center for Biotechnology Information. Subsequently, there were two federal investigations of my mistreatment, one by the U.S. Office of Special Counsel in 2005, and the other by subcommittee staff of the U.S. House Committee on Government Reform in 2006. Both investigations unearthed clear evidence that my rights had been repeatedly violated. Because there has been so much misinformation spread about what actually happened to me, I have decided to make available the relevant documents here for those who would like to know the truth."

See more at:

http://www.richardsternberg.com/smithsonian.php

https://answersingenesis.org/culture/the-smithsonian-sternberg-controversy/

http://blog.drwile.com/real-science-might-be-alive-and-well/

https://www.bing.com/search?q=Sternberg+peer+review+controversy&qs=SW&cvid=7d8c8af7f40542b4b05f4e5d772af1b5&pq=Sternberg+peer+review+controversy&cc=US&setlang=en-US&nclid=616CDAFDCB493AF7017B76771716BC56&ts=1536677593055&nclidts=1536677593&tsms=055&first=11&FORM=PORE

https://en.wikipedia.org/wiki/Sternberg_peer_review_controversy

And the Washington Post weighs in:

http://www.washingtonpost.com/wp-dyn/content/article/2005/08/18/AR2005081801680.html?noredirect=on

Evolutionary biologist Richard Sternberg made a fateful decision a year ago.

As editor of the hitherto obscure Proceedings of the Biological Society of Washington, Sternberg decided to publish a paper making the case for "intelligent design," a controversial theory that holds that the machinery of life is so complex as to require the hand -- subtle or not -- of an intelligent creator.

Within hours of publication, senior scientists at the Smithsonian Institution -- which has helped fund and run the journal -- lashed out at Sternberg as a shoddy scientist and a closet Bible thumper.

"They were saying I accepted money under the table, that I was a crypto-priest, that I was a sleeper cell operative for the creationists," said Steinberg, 42, who is a Smithsonian research associate. "I was basically run out of there."

An independent agency has come to the same conclusion, accusing top scientists at the Smithsonian's National Museum of Natural History of retaliating against Sternberg by investigating his religion and smearing him as a "creationist."

The U.S. Office of Special Counsel, which was established to protect federal employees from reprisals, examined e-mail traffic from these scientists and noted that "retaliation came in many forms . . . misinformation was disseminated through the Smithsonian Institution and to outside sources. The allegations against you were later determined to be false."

"The rumor mill became so infected," James McVay, the principal legal adviser in the Office of Special Counsel, wrote to Sternberg, "that one of your colleagues had to circulate [your résumé] simply to dispel the rumor that you were not a scientist."

The Washington Post and two other media outlets obtained a copy of the still-private report.

McVay, who is a political appointee of the Bush administration, acknowledged in the report that a fuller response from the Smithsonian might have tempered his conclusions. As Sternberg is not a Smithsonian employee -- the National Institutes of Health pays his salary -- the special counsel lacks the power to impose a legal remedy.

A spokeswoman for the Smithsonian Institution declined comment, noting that it has not received McVay's report.

"We do stand by evolution -- we are a scientific organization," said Linda St. Thomas, the spokeswoman. An official privately suggested that McVay might want to embarrass the institution.

It is hard to overstate the passions fired by the debate over intelligent design. President Bush recently said that schoolchildren should learn about the theory alongside Darwin's theory of evolution -- a view that goes beyond even the stance of intelligent design advocates. Dozens of state school boards have attempted to mandate the teaching of anti-Darwinian theories.

Richard Sternberg – the rest of the story.

But all this did not finish off Dr. Sternberg as a distinguished evolutionary biologist. He went on to join the ENCODE project, (acronym for ENCyclopedia Of DNA Elements) consisting of 400 scientists in 32 different laboratories studying "junk" DNA, the so-called "junk DNA" hypothesis. Mainstream evolutionary biologists such as Richard Dawkins have maintained, as a core evolutionary "fact" (one of the many facts in the mountain of evidence for evolution) that the majority of DNA is non-functional junk carried on as debris in the long history of gradual biological change. ENCODE turned that idea on its head when they found that some 80% of DNA has viable and necessary function. Some of the researchers speculate that eventually it will be found that 100% will have some functional purpose.

As part of this ENCODE project, Richard Sternberg looked at a specific type of junk DNA called retroelements. He produced several tables in his paper showing the function of many of them (mainly in regulation). He said, "*Our expectation is that one day, we will think of what used to be called 'junk DNA' as a critical component of truly 'expert' cellular control regimes.*"

The ENCODE project went against the scientific orthodoxy and examined "junk" DNA with a very unorthodox mission of following the evidence to wherever it may lead.

Remember the words of Dr. Richard Leontine " … we are forced by our a priori adherence to material causes to create an apparatus of investigation and a set of concepts that produce material explanations, no matter how counter-intuitive, no matter how mystifying to the uninitiated.

Moreover, that materialism is absolute, for we cannot allow a Divine Foot in the door … "

The Case of Eric Hedin and Ball State University

The following is the decision rendered by Dr. Jo Ann M. Gora of Ball State University regarding the case of Dr. Eric Hedin's courses *Inquiries in Physical Sciences* and *Astronomy 151 The Boundaries of Science* where Dr. Hedin introduces Intelligent Design in an elective graduate honors course. I've written about this case before, including Materialists (Atheists) Challenge A Course at Ball State University Which Introduces Intelligent Design.

Dear Faculty and Staff,

This summer, the university has received significant media attention over the issue of teaching intelligent design in the science classroom. As we turn our attention to final preparations for a new academic year, I want to be clear about the university's position on the questions these stories have raised. Let me emphasize that my comments are focused on what is appropriate in a public university classroom, not on the personal beliefs of faculty members.

Intelligent design is overwhelmingly deemed by the scientific community as a religious belief and not a scientific theory. Therefore, intelligent design is not appropriate content for science courses. The gravity of this issue and the level of concern among scientists are demonstrated by more than 80 national and state scientific societies' independent statements that intelligent design and creation science do not qualify as science. The list includes societies such as the National Academy of Sciences, the American Association for the Advancement of Science, the American Astronomical Society, and the American Physical Society.

Discussions of intelligent design and creation science can have their place at Ball State in humanities or social science courses. However, even in such contexts, faculty must avoid endorsing one point of view over others. The American Academy of Religion draws this distinction most clearly:

Creation science and intelligent design represent worldviews that fall outside of the realm of science that is defined as (and limited to) a method of inquiry based on gathering observable and measurable evidence subject to specific principles of reasoning. Creation science, intelligent design, and other worldviews that focus on speculation regarding the origins of life represent another important and relevant form of human inquiry that is appropriately studied in literature and social science courses. Such study, however, must include a diversity of worldviews representing a variety of religious and philosophical perspectives and must avoid privileging one view as more legitimate than others.

Teaching religious ideas in a science course is clearly not appropriate. Each professor has the responsibility to assign course materials and teach content in a manner consistent with the course description, curriculum, and relevant discipline. We are compelled to do so not only by the ethics of academic integrity but also by the best standards of our disciplines.

As this coverage has unfolded, some have asked if teaching intelligent design in a science course is a matter of academic freedom. On this point, I want to be very clear. Teaching intelligent design as a scientific theory is not a matter of academic freedom – it is an issue of academic integrity. As I noted, the scientific community has overwhelmingly rejected intelligent design as a scientific theory. Therefore, it does not represent the best standards of the discipline as determined by the scholars of those disciplines. Said simply, to allow intelligent design to be presented to science students as a valid scientific theory would violate the academic integrity of the course as it would fail to accurately represent the consensus of science scholars.

Courts that have considered intelligent design have concurred with the scientific community that it is a religious belief and not a scientific theory. As a public university, we have a constitutional obligation to maintain a clear separation between church and state. It is imperative that even when religious ideas are appropriately taught in humanities and social science courses, they must be discussed in comparison to each other, with no endorsement of one perspective over another.

These are extremely important issues. The trust and confidence of our students, the public, and the broader academic community are at stake. Our commitment to academic freedom is unflinching. However, it cannot be used as a shield to teach theories that have been rejected by the discipline under which a science course is taught. Our commitment to the best standards of each discipline being taught on this campus is equally unwavering. As I have said, this is an issue of academic integrity, not academic freedom. The best academic standards of the discipline must dictate course content.

Thank you for your attention to these important issues. Best wishes in your preparations for a new academic year. I look forward to seeing you at the fall convocation in just a few weeks.

Sincerely,

Jo Ann M. Gora, PhD President

My Disappointments in Dr. Gora's decision:

My discussions will be quite lengthy, so I will summarize here, and follow with detail. I hope you will give it some consideration.

1. The decision by Dr. Gora ignores, censors and disrespects the foundational history of modern science.
2. The decision misrepresents Intelligent Design as a new movement designed with the intention of somehow sneaking "creationism" into the public square.
3. The decision grossly misrepresents and exaggerates the success of Neo-Darwinian and its evidence, while at the same time dismissing the challenges of Intelligent Design and Creation Science.
4. The decision unnecessarily promotes an atheistic world view to students searching for answers and truth.
5. The decision totally closes out discussion and teaching of the history of science ... even in the "humanities and social science courses."
6. The decision continues the perversion of the 1'st Amendment to the Constitution.
7. The decision provides convenient cover and precedence for other academic institutions to follow suit.

The decision encourages a spirit of condescension and incivility towards people of faith.

And my follow up to the Hedin/Ball State censorship decision.

1. **The decision by Dr. Gora ignores, censors and disrespects the foundational history of modern science.**

"Intelligent design is overwhelmingly deemed by the scientific community as a religious belief and not a scientific theory. Therefore, intelligent design is not appropriate content for science courses."

The very foundation and origin of much of what is called modern science began with Bible believing Christian creationists. This science originated primarily in Europe in the 17'th, 18'th and 19'th centuries, well before Charles Darwin came on the scene. I base this position to a large extent on the following works, and encourage you to peruse them:

Christian Philosophy and the Origin of Science at: http://www.allaboutworldview.org/christian-philosophy-and-the-origin-of-science-faq.htm and: The Origin of Science at http://www.columbia.edu/cu/augustine/a/science_origin.html

And the book: Men of Science Men of God by Henry Morris.

The contributions of these scientific pioneers are impressive as shown below:

DISIPLINE	SCIENTIST
Antiseptic Surgery	Joseph Lister (1827-1912)
Bacteriology	Louis Pasteur (1822-1895)
Calculus	Isaac Newton (1642-1727)
Celestial Mechanics	Johann Kepler (1571-1630)
Chemistry	Robert Boyle (1627-1691)
Comparative Anatomy	Georges Cuvier (1769-1832)
Computer Science	Charles Babbage (1792-1871)
Dimensional Analysis	Lord Rayleigh (1842-1919)

Dynamics	Isaac Newton (1642-1727)
Electronics	John Ambrose Fleming (1849-1945)
Electrodynamics	James Clark Maxwell (1831-1879)
Electromagnetics	Michael Faraday (1791-1867)
Energetics	Lord Kelvin (1824-1907)
Entomology of Living Insects	Henri Fabre (1823-1915)
Field Theory	Michael Faraday (1791-1867)
Fluid Mechanics	George Stokes (1819-1903)
Galactic Astronomy	William Herschel (1738-1822)
Gas Dynamics	Robert Boyle (1627-1691)
Genetics	Gregor Mendel (1822-1884)
Glacial Geology	Louis Agassiz (1807-1873)
Gynecology	James Simpson (1811-1870)
Hydraulics	Leonardo da Vinci (1452-1519)
Hydrography	Matthew Maury (1806-1873)
Hydrostatics	Blaise Pascal (1623-1662)
Ichthyology	Louis Agassiz (1807-1873)
Isotopic Chemistry	William Ramsay (1852-1916)
Model Analysis	Lord Rayleigh (1842-1919)
Natural History	John Ray (1627-1705)
Non-Euclidian Geometry	Bernhard Riemann (1826-1866)

Oceanography	Matthew Maury (1806-1873)
Optical Mineralogy	David Brewster (1781-1868)
Paleontology	John Woodward (1655-1728)
Pathology	Ralph Virchow (1821-1902)
Physical Astronomy	Johann Kepler (1571-1630)
Reversible Thermodynamics	James Joule (1818-1889)
Statistical Thermodynamics	James Clerk Maxwell (1831-1879)
Stratigraphy	Nicholas Steno (1631-1686)
Systematic Biology	Carolus Linnaeus (1707-1778)
Thermodynamics	Lord Kelvin (1824-1907)
Thermo kinetics	Humphrey Davy (1778-1829)
Vertebrate Paleontology	George Cuvier (1769-1832)

And pay attention here to some notable inventions, discoveries, or developments by Bible-Believing scientists:

CONTRIBUTIONS	SCIENTIST
Absolute Temperature Scale	Lord Kelvin (1824-1907)
Actuarial Tables	Charles Babbage (1792-1871)
Barometer	Blaise Pascal (1623-1662)
Biogenesis Law	Louis Pasteur (1822-1895)
Calculating Machine	Charles Babbage (1792-1871)
Chloroform	James Simpson (1811-1870)

Classification System	Carolus Linnaeus (1707-1778)
Double Stars	William Herschel (1738-1822)
Electric Generator	Michael Faraday (1791-1867)
Electric Motor	Joseph Henry (1797-1878)
Ephemeris Tables	Johann Kepler (1571-1630)
Fermentation Control	Lois Pasteur (1822-1895)
Galvanometer	Joseph Henry (1797-1878)
Global Star Catalog	John Herschel (1792-1891)
Inert Gases	William Ramsay (1852-1916)
Kaleidoscope	David Brewster (1781-1868)
Law of Gravity	Isaac Newton (1642-1727)
Mine Safety Lamp	Humphrey Davy (1778-1829)
Pasteurization	Louis Pasteur (1822-1895)
Reflecting Telescope	Isaac Newton (1642-1727)
Scientific Method	Francis Bacon (1561-1626)
Self Induction	Joseph Henry
Telegraph	Samuel F. B. Morse (1791-1872)
Thermionic Valve	Ambrose Fleming (1849-1945)
Transatlantic Cable	Lord Kelvin (1824-1907)

Vaccination & Immunization — Louis Pasteur (1822-1895)

And from the paper <u>Christian Philosophy and the Origin of Science</u> we see:

An examination of the history of modern science reaffirms the supernaturalist's premise that science is not hostile to the Christian position. Modern science was founded by those who viewed the world from a Christian perspective. Francis Schaeffer writes,

"Since the world had been created by a reasonable God, scientists were not surprised to find a correlation between themselves as observers and the thing observed — that is, between subject and object. Without this foundation, modern Western science would not have been born."

Christianity was "the mother of modern science." Norman L. Geisler and J. Kerby Anderson's *Origin Science* contains a chapter titled "The Supernatural Roots of Modern Science." Both Alfred North Whitehead and J. Robert Oppenheimer defended this view. Philosopher and historian of science Stanley L. Jaki notes that historically the belief in creation and the Creator was the moment of truth for science: *"This belief formed the bedrock on which science rose."* Jaki powerfully defends this position in the *Origin of Science* and the *Savior of Science*. Rodney Stark comes to the same conclusion.

This shows that the modern Creation Science movement as founded by Henry Morris at the Institute for Creation Research, and the Intelligent Design (ID) movement as exemplified by the Discovery Institute are following the scientific tradition of those great scientists listed above.

Note also that a fundamental ID premise is that intelligence can be inferred and discovered without a dependence on religious texts such as the Bible.

* * * *

2. The decision misrepresents Intelligent Design as a new movement designed with the intention of somehow sneaking "creationism" into the public square.

I add to my argument in the previous 'disappointment' the following from the paper: "*How is it that science became a self-sustaining enterprise only in the Christian West?*"

... as Whitehead pointed out, it is no coincidence that science sprang, not from Ionian metaphysics, not from the Brahmin-Buddhist-Taoist East, not from the Egyptian-Mayan astrological South, but from the heart of the Christian West, that although Galileo fell out with the Church, he would hardly have taken so much trouble studying Jupiter and dropping objects from towers if the reality and value and order of things had not first been conferred by belief in the Incarnation. (Walker Percy, Lost in the Cosmos)

To the popular mind, science is completely inimical to religion: science embraces facts and evidence while religion professes blind faith. Like many simplistic popular notions, this view is mistaken. Modern science is not only compatible with Christianity, but it in fact also finds its origins in Christianity. This is not to say that the Bible is a science textbook that contains raw scientific truths, as some evangelical Christians would have us believe. The Christian faith contains deeper truths– truths with philosophical consequences that make conceivable the mind's exploration of nature: man's place in God's creation, who God is and how he freely created a cosmos.

In large part, the modern mind thinks little of these notions in much the same way that the last thing on a fish's mind is the water it breathes. It is difficult for those raised in a scientific world to appreciate the plight of the ancient mind trapped within an eternal and arbitrary world. It is difficult for those raised in a post-Christian world to appreciate the radical novelty and liberation Christian ideas presented to the ancient mind.

I would add to this my own speculative question ... Would the modern Neo-Darwinian Atheists have accomplished anywhere near what those Christians listed above accomplished?

Again, I say, this shows that the modern Creation Science movement as founded by Henry Morris at the Institute for Creation Research, and the Intelligent Design (ID) movement as exemplified by the Discovery Institute are following the scientific tradition of those great scientists listed above.

* * * *

3. The decision grossly misrepresents and exaggerates the success of Neo-Darwinian and its evidence, while at the same time dismissing the challenges of Intelligent Design and Creation Science.

We often hear of the 'overwhelming' evidence supporting Neo-Darwinian science, but when you start looking for it you see that it is to a large degree smoke and mirrors.

Example: Richard Dawkins book *'The Blind Watchmaker: Why the Evidence of Evolution Reveals a Universe without Design.'* I've looked for the evidence, and have found it to be the atheistic musings and philosophy in the mind of Dawkins; little more.

Example: Recent DNA studies show that at least 80% have functional value (predicted by ID) and are not 'Junk DNA' previously claimed as evolutionary evidence of natural selection discards.

Example: Scientists have recently loaded and retrieved the equivalent of 1,000,000 DVDs (music, and text) onto a small piece of DNA, showing the extreme complexity and information capability of DNA. Dawkins' response: *'Given infinite time, or infinite opportunities, anything is possible.'* Stephen Hawking's response: *'an undiscovered and undiscoverable multiverse offering an infinite combination of physical laws and constants, which ultimately and invariably will create DNA.'*

This DNA, by the way, is not just a collection of information, but carries with it the very plans, and even the mechanisms for constructing a large variety of life forms. Plans and mechanisms that for example will direct particular cells, at the proper times, to become: bone, skin, brain, muscle etc. Darwinists have no clue as to how or why this occurs, only the vague hope that someday science will provide the answers. ID proponents and Creationists on the other hand attribute this amazing device to be the product of 'mind.'

Example: The known laws of physics and the 230+ physical/chemical constants point to a universe fine-tuned to support life. Darwinian response: again, a multi-verse where anything can and will happen; no evidence, but plenty of theory.

Example: The fruit fly has been that subject of intense study for over 100 years, primarily because they are relatively simple and have very short generational life cycles allowing study across many generations. The poor creature has suffered numerous experiments in attempts to study Darwinian mutations. To date, 100% of these chemical, radiation and genetic manipulations have produced … defective fruit flies. I'm sure there have been (perhaps many) valid scientific discoveries and benefits from this 100+ years of study, but they have not produced an improved fruit fly, let alone a new species.

A great irony (at least in my mind) is that Professor Jerry Coyne, one of the most vocal proponents of Darwinism and one of the leading critics of Intelligent Design, has dedicated much of his own life in studying these little creatures; might he be continuing the efforts to positively demonstrate the power of Darwinism in creating a new species? Captain Ahab comes to mind.

Example: [Could the eye have evolved by natural selection in a geological blink?](#)

A search on the question above shows a refutation of the Darwinian claim that the eye could have evolved by Natural Selection, and I will let it speak for itself. Now let's complete the picture and show that the eye is but one important part of a very complex machine called the human body that exhibits the qualities of design that allow, for example, the execution of the complex double-play in baseball. So, let us for a moment consider only the head and its array of complex instruments.

My understanding is that the vision experience is a distributed process, distributed between the eye, the optic nerves, and the brain. At what point did the primitive creature decide it needed to distribute his vision? Did the eye, the optic nerves and the corresponding brain cells, along with the containing scull and controlling muscles evolve concurrently? In order to produce the fully functional vision experience the whole system most certainly would have to have been in synchronous evolution.

Evolutionary theory and literature speak of an incremental series of evolutionary steps improving the eye from a flat photosensitive spot to a full functioning eye such as you and I have, but makes no mention of the complete visual system ... only the eye.

At what point and how did the primitive creature decide it needed stereoscopic sight, a nice addition but not really necessary?

And did the paper address the various kinds of vision that the visual system processes? Items such as color differentiation, motion in all directions and speeds; intensity of the light?

The evolutionary development of the visual experience can certainly be imagined, but the claim that a fully developed vertebrate eye could have developed from a simple light-sensitive spot by a process of unguided natural selection, in "less than 364,000 years" is speculation and imagination, not part of the supposed "overwhelming mountain of evidence" supporting Darwinian evolution.

Might not the visual experience be more likely the result of an intelligent design?

I won't go into as much detail for the other data gathering instruments and data processing systems contained within the human skull; the ears, the nose, taste, the strap-down inertial reference system within the inner ear providing a sense of three-dimensional place and attitude (i.e., up/down, east/west/north/south, all axis acceleration). Then there is the brain itself which makes sense out of all this input creating a near complete multi-sensual experience we call life.

As scientists discover more and more about the characteristics of DNA, we are seeing that it contains the detailed plans for all this stuff inside our scull. Furthermore, we are seeing that DNA actually contains the mechanisms for constructing this amazing array of instrumentation within our scull. This is indeed an "ahh hah!" moment.

Might not all of this be more likely the result of an intelligent design?

* * * *

4. The decision unnecessarily promotes an atheistic world view to students searching for answers.

Once you have damaged or destroyed a student's faith in a divine creation as depicted in Genesis, Psalms, the Gospel of John and elsewhere, or destroying the idea of intelligent design in nature, it opens up fertile ground for a full-frontal attack on all the alleged evils of religion (read: Christianity and Judaism). And this attack leaves little room for examination of the very real and remarkable positive contributions of Christianity and Judaism to civilization:

This excerpt from Dr. D. James Kennedy's book: <u>What If Jesus Had Never Been Born?</u>

- Hospitals, which essentially began during the Middle Ages.
- Universities, which also began during the Middle Ages. In addition, most of the world's greatest universities were started by Christians for Christian purposes.
- Literacy and education for the masses.
- Capitalism and free-enterprise.
- Representative government, particularly as it has been seen in the American experiment.
- The separation of political powers.
- Civil liberties.
- The abolition of slavery, both in antiquity and in more modern times.

Modern science.

- The discovery of the New World by Christopher Columbus.
- The elevation of women.
- Benevolence and charity; the good Samaritan ethic.
- Higher standards of justice.
- The elevation of the common man.

- The condemnation of adultery, homosexuality, and other sexual perversions. This has helped to preserve the human race and has spared many from heartache.
- High regard for human life.
- The civilization of many barbarian and primitive cultures.
- The codifying and setting to writing of many of the world's languages.
- Greater development of art and music. The inspiration for the greatest works of art.
- The countless changed lives transformed from liabilities into assets to society because of the gospel.
- The eternal salvation of countless souls.

And this excerpt from Dr. D. James Kennedy's book: What If The Bible Had Never Been Written?

"When a young lady who was not a Christian heard about this book, 'What If the Bible Had Never Been Written?', she immediately said, "Oh the Bible has been nothing but oppressive towards women." This sentiment is often repeated in our biblically illiterate times. The truth is, the Bible has improved the treatment of women. Show me a country where women are treated well where the Bible has not gone first. You can't because it doesn't exist. In fact, chivalry — where women became protected and cherished – was started by the Church in the Middle Ages. When hundreds of men on the Titanic voluntarily gave up their lives so that women and children could use the lifeboats, they were following a centuries-old, cultural norm that the Bible had established. However, since the Bible has lost sway among many people in our culture today, I daresay if the Titanic were to sink now, I doubt if most men would so readily give up on trying to get into a lifeboat.

The young lady's opposition to the Bible (which she had never bothered to read) is far to typical today. It comes from the school of ignorant thought that says Christianity is sexist, homophobic, racist, anti-science, anti-progress, and several other negative things.

What if the Bible had never been written? Consider the implications of such a scenario. There would be no salvation, no Salvation Army, no YMCA, virtually no charity, no modern science, no Red Cross. There would likely be no hospitals, for hospitals, as we know them, were born in the Christian era, and Christians have built hundreds of hospitals all over the globe. There would probably be no universities; they were created in the Middle Ages in order to reconcile Christian theology with the writings of Aristotle. There would probably be no capitalism, no accounting, no free enterprise. Millions of people would have been killed off by STDs (sexually transmitted diseases) – without any kind of inhibition against sexual promiscuity. Literacy and education might well have been the exclusive domain of the elite. Many of the languages around the world would never have been written down because there would have been no motive to do so. Many of the barbarians of the world over would never have been civilized. Cannibalism and human sacrifice and the abandonment of children would still be widespread, even as abortion and infanticide plague us as we continue to move away from the Bible. Slavery might still be practiced, as it is in pockets of the world where the Bible is forbidden. And we might not even be in the New World – as Columbus clearly stated it was the Lord who inspired him to make his historic voyage. If the Bible had never been written, there would be no Mother Teresa, no David Livingstone, no Isaac Newton, no William Wilberforce, no George Washington, no Abraham Lincoln, no Dante, no Milton, no Shakespeare, no Dickenson. Above all, if the Bible had never been written, we would be cut off from God, groping in darkness without hope.

But the Bible has been written, and we can embrace its wonderful message of the love of God, which is so great that He gave His only begotten Son that we may have eternal life. Because the Bible has been written, the wonderful story of how Jesus came to seek and save the lost has gone out into all the world and has transformed millions of lives and scores of cultures and nations."

* * * *

5. The decision continues the perversion of the 1'st Amendment to the Constitution.

The so called *"separation of church and state"* is often abused and in fact does not appear in the 1'st Amendment which reads in part:

"Congress shall make no law respecting an establishment of religion, or prohibiting the free exercise thereof ... "

A more historically accurate paraphrase would be *"separation of church from the state."* Rendering in this fashion captures the founder's intention of keeping the national government out of the business of the church, while at the same time allowing the state to be influenced, but not controlled, by the church.

Further, it is indeed a tortured stretch to claim that Congress has somehow stepped into a classroom in Muncie Indiana and established a state religion there.

* * * *

6. The decision totally closes out discussion and teaching of the history of science ... even in the "humanities and social science courses."

Dr. Gora's missive to Faculty and Staff states forcefully: *" ... It is imperative that even when religious ideas are appropriately taught in humanities and social science courses, they must be discussed in comparison to each other, with no endorsement of one perspective over another."*

But this, along with the previous ruling regarding science classes, places the University, faculty and students in an untenable box. That box being ... how is the *'History of Science'* to be taught?

It can't be taught accurately as a science course because Kepler (1571-1630) stated that he was merely *"thinking Gods thoughts after Him,"* a motto adopted by many believing scientists of the day.

Or Francis Bacon (1561 -1626), considered to be the man primarily responsible for the formulation and establishment of the so-called "scientific method." He wrote *"There are two books laid out before us to study, to prevent our falling into error; first, the volume of the scriptures, which reveal the will of God; then the volume of the Creatures, which express His power."*

There are other such testimonies of the men listed in the tables above, but I'm afraid their stories can't be told with any accuracy at Ball State University.

You see, they were Intelligent Design proponents who believed mind ... the mind of God ... intelligent design ... was behind what they saw around them in the world, and above them in the heavens. The mind of God, and his power and intelligence are what motivated these scientists.

So ... out goes History of Science, at least in the science curriculum.

Well then, that leaves "humanities and social science courses" as long it is "discussed in comparison to each other, with no endorsement of one perspective over another."

But here is the problem:

How Christianity Kick-Started Modern Science | Catholic Answers Magazine ...as Whitehead pointed out, it is no coincidence that science sprang, not from Ionian metaphysics, not from the Brahmin-Buddhist-Taoist East, not from the Egyptian-Mayan astrological South, but from the heart of the Christian West, that although Galileo fell out with the Church, he would hardly have taken so much trouble studying Jupiter and dropping objects from towers if the reality and value and order of things had not first been conferred by belief in the Incarnation. (Walker Percy, *Lost in the Cosmos*)

Thus we arrive at the box (or cave) containing Ball State University.

It would seem the History of Science can only be taught at BSU by being Politically Correct with regard to non-European, and non-Christian civilizations. This could be accomplished easily in one of two ways; by referring to Newton and others as simply members of the human family with no further discriminators; or by teaching that science began with the arrival of Charles Darwin.

* * * *

7. The decision provides convenient cover and precedence for other academic institutions to follow suite.

No comment necessary.

8. The decision encourages a spirit of condescension and incivility towards people of faith.

Here's an example of what Dr. Jerry Coyne, an atheist and honorary board member of the atheistic Freedom From Religion Foundation; the leader in this action against Ball State, has to say:

"… If what makes you "you" is a belief in delusions, like your redemption through the execution of a Palestinian carpenter, or the notion that a cracker and wine literally—literally—become the body and blood of that carpenter, then you're fair game for criticism. And plenty of people think the core of their being rests on belief in the Genesis story of creation and a young earth, the idea of psychic phenomena, or their imagined abduction by aliens. Are we to coddle them as well?

None of these ideas deserve dignity. And the same Church that provides Stanley with compassion and friendship also marginalizes women, prohibits abortion, divorce, and gay behavior, terrorizes children with thoughts of hell, sanctions and protects child rape, and deliberately spreads AIDS in Africa by denying its adherents birth control. Are we to remain silent on these, too?"

I have been following the sort of comments generated by readers of Dr. Coyne's web site Why Evolution is True, and other sites, and they are truly full of condescension and incivility towards people of faith. And, dissenting comments such as mine, are trashed and never appear.

I have put my oar in the water in places like the Huffington Post which have generated responses such as:

"quibbling with the educated", "don't EVER quote ... It just makes your ignorance clear", "grow up", "ridiculous rhetoric.", "believing in an imaginary friend", "Ridiculous ideas deserve ridicule", "your statements are so ridiculous they deserve ridicule", "utterly ridiculous", "I'm assuming your[sic] just a little educated", "to get more money out of your wallet ", "your comment again deserves complete ridicule", "I hire and fire code monkeys ", "you are lacking education ... you are ignorant", "These are evil, immoral and dishonest men", "then you too are an evil, immoral and dishonest man. Period", and more.

I can imagine what a university student faces. Presented with only the atheistic materialistic/naturalistic world view with little or no contrary point of view, is there a fair amount of brow beating and condescension going on? My own experience has been that in the American culture, including and especially college, there is a fog of pervasive evolutionary thought. Counter it at your own peril.

A somewhat personal encounter with professional bigotry

I had a somewhat personal encounter with this sort of threat a few years ago.

While in Seattle, I had occasion to have lunch with Casey Luskin of the Discovery Institute. Along with Casey was a young man who as I recall was either in a PhD program or Post Doc somewhere. A very nice time was had with both.

A day or so following the lunch, I posted a report of that meeting to my blog, along with the name of the young man and where he was from. I forwarded a copy of that post to Casey.

Bad decision on my part. I got an urgent message from Casey as we were driving along a remote road west of Seattle. Casey was angry with me. He explained that by posting this personal information to a public blog, I was putting this young man's future career in jeopardy. On my cell phone on that remote road, I managed to delete my post. I subsequently did some research on people like Dr. Hedin, Dr. Sternberg and others whose career and livelihood were put in jeopardy because of their stance on Intelligent Design.

Here's what Sara Chaffee and Ann Gauger of the Discovery Institute had to say in very recent months:

"Yet as Ann Gauger pointed out here the other day, we can't even safely show you photos of the Seminars without carefully cropping out faces, even backs of heads! That's for fear that we will expose our Seminar graduates to retaliation when they return to their home campuses, or other career perils. This year's Seminar included some severe warnings to the group about how, as a promising young scholar with doubts about Darwinism, you need to protect yourself from such dangers."

https://evolutionnews.org/2018/07/evolution-and-the-i-wont-debate-tactic/

Conclusion

Much more could be written, but I conclude with this thought:

No, I am not an Evolutionary Biologist of any sort. Nor am I an automotive designer or engineer of any sort.

But ... I am able to judge that my 2016 Nissan Rouge is far superior in every way than my old 1949 Ford.

About the author

I was born and raised in Butte Montana, joined the Navy in 1964 and soon thereafter married Diana Lingley, my high school sweetheart. We raised two children and have three grandchildren.

After serving in the Navy, I returned to school in San Diego where I earned a bachelor's degree in mathematics and launched on a long career in software development.

This long term, mostly hands on career in software and systems has given me a strong sense of design and what it takes to develop and maintain very complex systems.

I have no advanced degrees in chemistry, or biology (evolutionary or otherwise) or microbiology, but as an interested layman have cultivated a long term, passionate interest in intelligent design/creation/evolution and the implications presented by each.

What I have hopefully honed over the years is heightened common sense and powers of observation and reason.

To those who would question my qualifications to weigh in on these issues, I would ask the following:

As a software developer, if I were to claim that software intensive systems such as Microsoft Windows arose by way of a bunch of geeks just hanging out for a very long time randomly plunking on a keyboard ... and if you questioned that claim, would you be justified in your criticism? I think you would. On the other hand, I would expect you to have a degree of expertise when criticizing and commenting on the merits of the .NET framework which is a vital part of the Microsoft software infrastructure.

As a consumer of automobiles as well as a consumer of ideas, but neither an automotive designer nor engineer, I am able to judge – as are you - that my near new Nissan Rouge is far superior to my old 1949 Ford.

What I would ask of my readers is that they do their own homework and don't blindly follow charismatic "experts." Learn from them and study them and gain wisdom from them but recognize when they drift away from their particular field of expertise into ideology and philosophy.

Some references to get you a start (far from exhaustive).

http://www.icr.org/topics/
http://whomadegod.org/
http://www.uncommondescent.com/
http://www.icr.org/
http://www.discovery.org/
http://www.answersingenesis.org/
http://whyevolutionistrue.wordpress.com/
http://sciencebasedlife.wordpress.com/
http://www.evolutionnews.org/
http://www.AYearningForPublius.Wordpress.com (My web site) In CATEGORIES, click on "Intelligent Design" and "Evolution"
http://NCSE.com

This book and others I have written may be purchased at:

amazon.com/author/donjohnsonbooks

and

http://www.blurb.com/b/8349209-sam-jankovich
Memoirs of a sports legend

Contact: DonJohnsonDD682@live.com

brain ... eyes ... ears ... fingers.

Illusions or real ... do they sense what's real or do they simply imagine?